Statistics for
Archaeologists

A Commonsense Approach

D0141327

INTERDISCIPLINARY CONTRIBUTIONS TO ARCHAEOLOGY

Series Editor: Michael Jochim, *University of California, Santa Barbara*
Founding Editor: Roy S. Dickens, Jr., *Late of University of North Carolina, Chapel Hill*

Current Volumes in This Series:

THE ARCHAEOLOGY OF WEALTH
Consumer Behavior in English America
James G. Gibb

CASE STUDIES IN ENVIRONMENTAL ARCHAEOLOGY
Edited by Elizabeth J. Reitz, Lee A. Newsom, and Sylvia J. Scudder

CHESAPEAKE PREHISTORY
Old Traditions, New Directions
Richard J. Dent, Jr.

DARWINIAN ARCHAEOLOGIES
Edited by Herbert Donald Graham Maschner

DIVERSITY AND COMPLEXITY IN PREHISTORIC MARITIME SOCIETIES
A Gulf of Maine Perspective
Bruce J. Bourque

HUMANS AT THE END OF THE ICE AGE
The Archaeology of the Pleistocene–Holocene Transition
Edited by Lawrence Guy Straus, Berit Valentin Eriksen, Jon M. Erlandson,
and David R. Yesner

PREHISTORIC CULTURAL ECOLOGY AND EVOLUTION
Insights from Southern Jordan
Donald O. Henry

REGIONAL APPROACHES TO MORTUARY ANALYSIS
Edited by Lane Anderson Beck

STATISTICS FOR ARCHAEOLOGISTS
A Commonsense Approach
Robert D. Drennan

STONE TOOLS
Theoretical Insights into Human Prehistory
Edited by George H. Odell

STYLE, SOCIETY, AND PERSON
Archaeological and Ethnological Perspectives
Edited by Christopher Carr and Jill E. Neitzel

A Chronological Listing of Volumes in this series appears at the back of this volume.

A Continuation Order Plan is available for this series. A continuation order will bring delivery
of each new volume immediately upon publication. Volumes are billed only upon actual
shipment. For further information please contact the publisher.

Statistics for Archaeologists

A Commonsense Approach

ROBERT D. DRENNAN

University of Pittsburgh
Pittsburgh, Pennsylvania

 Springer

Library of Congress Cataloging-in-Publication Data

Drennan, Robert D.
    Statistics for archaeologists : a commonsense approach / Robert D.
Drennan.
        p.   cm. -- (Interdisciplinary contributions to archaeology)
    Includes bibliographical references and index.
    ISBN 0-306-45327-4 (hardcover). -- ISBN 0-306-45326-6 (pbk.)
    1. Archaeology--Statistical methods.   I. Title.   II. Series.
CC80.6.D74   1996
930.1'021--dc20   96-17073
                                                                 CIP

Printed in the United States of America.

9 8 7 6 5

springeronline.com

Preface

This book is intended as an introduction to basic statistical principles and techniques for the archaeologist. It grows primarily from my experience in teaching courses in quantitative analysis for undergraduate and graduate students in archaeology over a number of years. The book is set specifically in the context of archaeology, not because the issues dealt with are uniquely archaeological in nature, but because many people find it much easier to understand quantitative analysis in a familiar context—one in which they can readily understand the nature of the data and the utility of the techniques. The principles and techniques, however, are all of much broader applicability. Physical anthropologists, cultural anthropologists, sociologists, psychologists, political scientists, and specialists in other fields make use of these same principles and techniques. The particular mix of topics, the relative emphasis given them, and the exact approach taken here, however, do reflect my own view of what is most useful in the analysis of specifically archaeological data.

It is impossible to fail to notice that many aspects of archaeological information are numerical and that archaeological analysis has an unavoidably quantitative component. Standard statistical approaches are commonly applied in straightforward as well as unusual and ingenious ways to archaeological problems, and new approaches have been invented to cope with the special quirks of archaeological analysis. The literature on quantitative analysis in archaeology has grown to prodigious size in the past 25 or 30 years. Some of this literature is extremely good, while some of it reveals only that publishing on statistics in archaeology is an activity open even to those whose comprehension of the most fundamental statistical principles is primitive at best. The article attempting to point out which published work fits into which of these categories has itself become a recognizable genre. This book does not attempt to evaluate or criticize in such a mode, but it is motivated in part by the perception that, as a group, those of us responsible for training archaeologists in quantitative analysis can claim only mixed suc-

v

cess to date. Consequently, this book is in part a discussion of how quantitative data analysis is done in archaeology but in larger part a discussion of how quantitative data analysis *could be* done in archaeology. Its focus is resolutely on some fundamental principles and how they can be applied most usefully in archaeology. It is tempting to discuss the numerous variations in these applications that might be made in analyzing archaeological data and to provide examples of ways in which these principles have actually been put to work by archaeologists. I have, however, attempted to resist these temptations in an effort to keep the focus firmly on basic principles and to provide brief and clear explanations of them. It is to maintain simplicity and clarity that both the examples used in the text and the practice problems at the ends of the chapters are made up rather than selected from real archaeological data. I assume that the readers of this book know enough about archaeology not to need descriptions and pictures of post holes, house floors, scrapers, or sherds—that we all know what it means to say that we have conducted a regional survey and measured the areas of 53 sites.

Most of the techniques in this book are fairly standard, either in the "classical" statistics developed between 1920 and 1950 or in the more recent "exploratory data analysis" school. The approach or, perhaps more important, the general attitude of this book derives ultimately from the work of John W. Tukey and his colleagues and students, progenitors of exploratory data analysis, or EDA for short. As is usual in general books on statistics, I have not included bibliographic citations in the text, but Suggested Reading appears at the end. This book leans toward the terminology of EDA, although the equivalent more traditional terms are usually mentioned. Where it makes the explanations easier to understand in the context of archaeology, the terminology used here is simply nonstandard.

Archaeologists (and others) sometimes are as wary of statistics as school children are of the classroom holding the most imposing disciplinarian among the teachers. Statistics seems a place filled with rules the rationale of which is opaque but the slightest infraction of which may bring a painful slap across the knuckles with a ruler. This attitude has certainly been reinforced by critiques that take published work in archaeology to task for breaking sacred statistical rules. It may come as a surprise to many to learn that a number of conflicting versions exist of many statistical rules. Statisticians, like the practitioners of any other discipline, often disagree about what are productive approaches and legitimate applications. Use of statistical tools often involves making subjective judgments. In an effort to provide a sound basis for such judgments, introductory texts often attempt to reduce them to clearcut rules, thereby creating considerable confusion about what are really fundamental principles and what are merely guidelines for difficult subjective decisions.

In short, the rules of statistics were not on the stone tablets Moses brought down from the mountain. This book openly advocates the overthrow of rules found in some texts (by reason and common sense rather than force and violence). Since it is intended as an introduction to statistical principles, long arguments against alternative approaches are not appropriate. One issue, however, is of such central importance that it must be mentioned. The approach taken to significance testing here does not involve rigid insistence on either rejecting or failing to reject a "null hypothesis." In archaeology it is much more informative in most instances simply to indicate how likely it is that the null hypothesis is correct. The rigorous formulation of the null hypothesis, then, does not get the all-consuming attention here that is sometimes devoted to it elsewhere. In this approach to significance testing and to several issues related to sampling, I have followed the lead of George Cowgill (see Suggested Reading at the end of the book), although I have not carried into practice all of the thoroughly sensible suggestions he has made. (One obstacle to following some of his suggestions continues to be, as he noted, that few of the available statistics computer programs provide the necessary information in their output.) To those who were taught that significance testing was built upon the rock of rejecting or failing to reject the null hypothesis, I recommend thoughtful attention to the points Cowgill makes.

The approach taken to significance testing makes clear thinking about populations, samples, and sampling procedures especially important. Indeed, in many contexts, it makes simply using samples to make statements about the populations they came from a more appealing approach than significance testing. It is for this reason that samples and sampling are given much lengthier treatment here than is common in introductory books on statistics. Part I of this book is about exploring batches of numbers in ways that are interesting and useful in and of themselves, but that are especially chosen for their relevance when batches are considered samples from larger populations. Part II develops this notion of batches as samples and makes a frontal assault on some of the central principles that relate samples to populations. Part III presents a fairly standard suite of basic tests of the strength and significance of relationships between two variables, together with alternative approaches derived directly from sampling estimation. Part IV returns to take up a series of separate issues related to sampling—issues of special importance in archaeology. These chapters relate most directly to those in Part II, but they have been placed at the end of the book to avoid interrupting the steady progression of ideas that links Parts II and III.

In archaeology, as in most fields, quantitative concepts come easily and naturally to some, and only at considerable cost to others. The absence of a natural inclination toward numerical reasoning is often reinforced by the social acceptability of professing ignorance of mathematics—a social acceptability nurtured by the notion that mathematics is an arcane and specialized subject of no use to very many people. An otherwise well-educated person can profess a complete inability to comprehend anything about numbers beyond addition and subtraction without incurring the disdain to be expected if he or she admitted to verbal skills so limited as to make everything in the daily newspaper but the comics unintelligible.

Varying degrees of natural talent should be no more surprising for mathematics than for writing, playing football, or other activities. The view that mathematics is only a necessary evil of elementary school, however, aggravates the problem by encouraging those who have found quantitative reasoning difficult to minimize its importance and to avoid developing quantitative skills that would be useful to them. Consequently, a good many students seem to embark on graduate study of archaeology equipped only with high school algebra—victims, perhaps, of the same kind of bad advice I myself received as a first-semester freshman in college, when my academic advisor scornfully dismissed the math course I intended to enroll in as irrelevant to my interests.

This book is written in the hope of providing useful tools for quantitative analysis in archaeology to those naturally adept at quantitative reasoning as well as to those who find mathematics not only difficult but even intimidating. It is no challenge to present statistics to those already comfortable with and adept at mathematical thinking; it requires only a nudge in the right direction. The perennial challenge of books such as this, however, is to present quantitative analysis effectively to those to whom it does not come naturally. It is with particular concern for this latter group that the approach taken here was chosen. Part of that approach is to plunge right ahead to the tools this book is about without a series of preliminary chapters laying basic groundwork, the importance of which only becomes apparent later on. These "basics" are, instead, discussed as briefly as possible at the points where they become relevant.

Fortunately, it is possible to approach basic statistical tools with common sense and in common language so as to convey not only the mechanics of using the tools of statistics but also a genuine understanding of the way the tools work. Productive use of statistical tools in archaeology springs not so much from abstract mathematical knowledge as from solid intuitive understanding of principles, applied with common sense and unwavering attention to the final product desired—that is, the ultimate research objective. It is worth pausing to emphasize that this book, fundamentally, is about tools—tools for identifying patterns in numbers and tools for assessing how precisely and how reliably the patterns we identify in our data represent real patterns in the broader world our conclusions really are about. As with carpenters' tools, for example, skillful use of statistical tools does not require complete knowledge of how the tools are made. Consequently, I have not attempted to show how statistical equations are derived from certain assumptions through mathematical logic (the approach followed by some books on statistics). As powerful and elegant as the language of abstract mathematics may be, it remains utterly impenetrable to many archaeologists. I have always found it helpful to avoid an abstract mathematical approach. This seems especially important to those already frightened at the thought of mathematics.

While learning to use a table saw does not require developing the ability to make one, skillful use of a table saw does require some understanding of the principles according to which it does its work. Failure to understand these basic principles will lead to erroneous and uneven cutting and even the occasional severed finger or worse. In just the same way, skillful use of statistical tools requires true understanding of underlying principles. Without such understanding, even very keen statistical tools produce only crude results, and they can cause injury (although generally not the kind that requires medical attention).

For this reason, I have also tried to avoid the cookbook approach common to books on applied statistics. Easy recipes for statistical analysis appeal strongly, especially to those afraid of mathematics. No real mental labor seems to be required, no difficult concepts need be mastered; just carefully follow the instructions. This approach may actually work in disciplines where certain kinds of data are regularly produced in certain formats. Only the most routine data analysis tasks can be successfully handled in this manner, however, and archaeological data are never routine. The nature of the archaeological record and the manner in which we must extract data from it inevitably produce idiosyncrasies that practitioners in other disciplines are taught to avoid through appropriate research design. Coping with such messy data requires that the archaeologist have a better grasp of underlying principles than a cookbook approach can provide.

This book, then, seeks a middle ground. It attempts more than simply providing instructions for the use of statistical tools, yet it makes no pretense to provide a complete mathematical justification for them. Its aim is to help the reader understand the principles underlying statistical tools well enough to use them skillfully in the context of archaeological data analysis. The reader I had in mind while writing is primarily the graduate or undergraduate student of archaeology taking a first course in archaeological data analysis. Like most textbooks, this is the book the author has always wanted but never found for his own course. I hope it may also be useful to archaeologists who wish to develop or consolidate skills in statistical tool use whether they are enrolled in courses or not.

The statistical tools discussed in this book by no means make up the complete set ever needed by the archaeologist. They are basic general-purpose tools, but many other specialized tools exist. Some of the tools presented here are quite simple and easy to apply, requiring nothing more than

pencil and paper or perhaps an ordinary calculator. Others are more complicated or involve very cumbersome calculations. I take it for granted that any
serious archaeological data analysis effort will now be undertaken with the
aid of a computer. Learning to use statistical software packages is best incor-
porated directly into the process of learning about the statistical tools. I have
thus omitted the often time-consuming and complex explanations of how to
compute certain complicated statistics by hand. While calculating some
things out by hand can facilitate understanding, one soon reaches the point
where preoccupation with the mechanics of calculations interferes with attention that should be devoted directly to underlying principles.

Many of the results and examples in this book were produced with
SYSTAT®; other packages that could be used are too numerous even to list.
Since the possibilities are so varied (and change so continually), it is useless
to attempt to incorporate instructions for using statistical software into this
book. I assume, however, that the book will be used in conjunction with
some package of statistical programs and the corresponding manuals, and
some general comments about using such "statpacks" are included.

Almost any software package will provide options and choices not discussed in this book. Some software manuals provide good explanations of
what these options are and bibliographic citations for those interested in
learning more about them; other manuals do not. (This is one feature worth
weighing in choosing statistical software.) Serendipitous encounters with
options in statistical software can provide a useful means of expanding one's
expertise in quantitative analysis. On the other hand, they can distract the
analyst's attention from the task at hand to the many other tasks that could
be performed but that there is really no need to perform. The professional
carpenter does not first choose a pretty tool and then go looking for something to use it on. Just so, the skilled data analyst first determines what
analysis to perform and then turns to pencil, paper, calculator, or computer
(as may be appropriate) to put into use the appropriate tool to accomplish
the task at hand. Both the mechanics of complicated calculations and complicated computer software can divert attention away from central matters of
principle concerning the work to be done. In statistics, as in the several
sports from which the cliché is derived, it is impossible to remind yourself
too often to keep your eye on the ball.

ACKNOWLEDGMENTS

The person most responsible for "infecting" me (his word, not mine)
with the attitude toward statistics represented here is Lee Sailer. Mark Aldenderfer and Doug Price provided very helpful reactions to the manuscript. I
have stubbornly refused to accept some of the advice generously offered by

all three, however, so they cannot be blamed for any deficiencies. Jeanne Ferrary Drennan has put up with a lot of cursing as I tried to teach courses in archaeological data analysis with texts I didn't like, and she dedicated most of one December vacation to helping put the first draft of this book in shape to use in class in January. My most special thanks, however, are reserved for the graduate and undergraduate students (and the teaching assistants) who have struggled gamely along as I tried to give enough coherence to this approach to data analysis in archaeology to use it in the courses they took—sometimes using texts it contradicted, sometimes using no text at all, and finally using successive draft versions of this book. They have contributed more than they know to whatever clarity the exposition here may have.

Robert D. Drennan

Pittsburgh, Pennsylvania

Contents

PART I. NUMERICAL EXPLORATION

PART IV. SPECIAL TOPICS IN SAMPLING

List of Reference Tables

Part I

Numerical Exploration

Chapter *1*

Batches of Numbers

A *batch* is a set of numbers that are related to each other because they are different instances of the same thing. The simplest example of a batch of numbers is a set of measurements of different examples of the same kind of thing. For example, the lengths of a group of scrapers, the diameters of a group of post holes, and the areas of a group of sites are three batches of numbers. In these instances, length, diameter, and area are *variables* and each scraper, post hole, and site is a *case*.

The length of one scraper, the diameter of one post hole, and the area of one site do not, together, make a batch of numbers, because they are completely unrelated. The length, width, thickness, and weight of one scraper do not, together, make a batch, because they are not different instances of the same thing; that is, they are different variables measured for a single case. The length, width, thickness, and weight of each of 20 scrapers make, not one batch of numbers, but four. These four batches can be related to each other because they are four variables measured for the same 20 cases. The diameters of a set of 18 post holes from one site and the diameters of a set of 23 post holes from another site can be considered a single batch of numbers (the variable diameter measured for 41 cases, ignoring entirely in which site each post hole appeared). They can also be considered two related batches of numbers (the variable diameter measured for 18 cases at one site and 23 cases at another site). Finally, they can be considered two related batches of numbers in a different way (the variable diameter measured for 41 cases and the variable site classified for the same 41 cases). This last, how-

ever, carries us to a different kind of batch or variable, and it is easier to stick
to batches of measurements for the moment.

STEM-AND-LEAF PLOTS

A list of measurements does not lend itself very well to making interesting observations, so the first step in exploration of a batch of numbers is to
organize them. If the batch is a set of measurements, the *stem-and-leaf plot* is
the fundamental organizational tool. Consider the batch of numbers in Table
1.1. It is difficult to say much about these measurements from a quick inspection of the table. Ordering them along a scale can often help us to see patterns.
Table 1.2 shows how to produce a stem-and-leaf plot that does exactly this for
the numbers in Table 1.1. First, the numbers are divided into a stem section
and a leaf section. In the first case, for instance, 9.7 becomes a stem of 9 and a
leaf of 7. The leaf for each number is placed on the stem plot beside the stem
for that number. The lines in Table 1.2 connect some of the numbers to the
corresponding leaves in their final positions on the stem-and-leaf plot. (To
avoid a hopeless confusion of lines, not all the connections are drawn in.)

Several characteristics of this batch of numbers are immediately apparent in the stem-and-leaf plot. First, the numbers tend to bunch together at
about 9 to 12 cm. Most fall in this range. Two more (14.2 cm and 7.6 cm) fall
a little outside this range, and one (44.6 cm) falls far away from the rest. It is
a fairly common occurrence for batches of numbers to bunch together like
this. It is also relatively common for one or a few numbers in a batch to fall
far away from the bunch where the majority of the numbers lie. Such numbers that fall far from the bunch are often called *outliers*, and we will discuss
them in more detail later. For now it is sufficient to note that we often
examine such outliers with a skeptical eye. A post hole 44.6 cm in diameter
is certainly a very unusual post hole in this batch, and we might be suspi-
cious that someone has simply written the measurement down wrong. A
quick check of field drawings or photographs should be sufficient to determine whether such an error has been made and, if so, to correct it. If, indeed,
this measurement seems correct, then one of the conspicuous features of this
batch is that one post hole simply does not seem to fit with the rest of the
group.

Table 1.1. Diameters of 13 Post Holes at the Black Site (cm)

9.7	9.1	11.1	10.8
9.2	44.6	7.6	
12.9	10.5	11.8	
11.4	11.7	14.2	

Table 1.2. A Stem-and-Leaf Plot of the Numbers in Table 1.1

STEMS	LEAVES		
		44	6
		43	
		42	
9.7	9	7	41
			40
9.2	9	2	39
			38
12.9	12	9	37
			36
11.4	11	4	35
			34
9.1	9	1	33
			32
44.6	44	6	31
			30
10.5	10	5	29
			28
11.7	11	7	27
			26
11.1	11	1	25
			24
7.6	7	6	23
			22
11.8	11	8	21
			20
14.2	14	2	19
			18
10.8	10	8	17
			16
			15
			14	2
			13
			12	9
			11	1478
			10	58
			9	127
			8
			7	6

Stem-and-leaf plots can be made at different scales (that is, using different intervals on the stem), and the selection of an appropriate scale is essential to producing a helpful stem-and-leaf plot. Table 1.3 shows another batch of numbers in a stem-and-leaf plot at the same scale as in the previous example. The numbers here, however, are spread out over such a large distance that the characteristics of the batch are not clearly displayed. In Table 1.4 the same numbers yield a denser stem-and-leaf plot when the stem is structured differently. In the first place, the numbers are broken differently into stem and leaf sections—not at the decimal point but between the units and tens. Since there are two digits for each leaf, commas are used to indicate the separation between leaves. To avoid greatly increasing the density, two

Table 1.3. Too Sparse a Stem-and-Leaf Plot of Weights of 17 Scrapers from the Black Site

Weight (g)	Stems	Leaves
148.7	148	7
154.5	154	5
169.5	169	5
145.1	145	1
157.9	157	9
137.8	137	8
151.9	151	9
146.2	146	2
164.7	164	7
149.3	149	3
141.3	141	3
161.2	161	2
146.9	146	9
152.0	152	0
143.0	143	0
132.6	132	6
115.3	115	3

```
169 | 5
168 |
167 |
166 |
165 |
164 | 7
163 |
162 |
161 | 2
160 |
159 |
158 |
157 | 9
156 |
155 |
154 | 5
153 |
152 | 0
151 | 9
150 |
149 | 3
148 | 7
147 |
146 | 29
145 | 1
144 |
143 | 0
142 |
141 | 3
140 |
139 |
138 |
137 | 8
136 |
135 |
134 |
133 |
132 | 6
131 |
130 |
129 |
128 |
127 |
126 |
125 |
124 |
```

Table 1.3.　(*Continued*)

Weight (g)	Stems	Leaves
	123	
	122	
	121	
	120	
	119	
	118	
	117	
	116	
	115	3

positions are allowed on the stem for each stem section, the lower position corresponding to the lower half of the numbers that might fit that stem section and the upper corresponding to the upper half (as indicated by the notations to the right of the stem-and-leaf plot). The characteristics of the batch are much clearer in this plot. The numbers bunch together from about 130 to 160. And one unusually light scraper seems to be an outlier. This pattern can certainly be detected (especially in hindsight) in Table 1.3, but it is much clearer in Table 1.4.

Table 1.4.　Stem-and-Leaf Plot at an Appropriate Scale of Weights of 17 Scrapers from the Black Site

Weight (g)	Stems	Leaves	Stems	Leaves	
148.7	14	87			
154.5	15	45			
169.5	16	95	17		(175.0–179.9)
145.1	14	51	17		(170.0–174.9)
157.9	15	79	16	95	(165.0–169.9)
137.8	13	78	16	12, 47	(160.0–164.9)
151.9	15	19	15	79	(155.0–159.9)
146.2	14	62	15	19, 20, 45	(150.0–154.9)
164.7	16	47	14	51, 62, 69, 87, 93	(145.0–149.9)
149.3	14	93	14	13, 30	(140.0–144.9)
141.3	14	13	13	78	(135.0–139.9)
161.2	16	12	13	26	(130.0–134.9)
146.9	14	69	12		(125.0–129.9)
152.0	15	20	12		(120.0–124.9)
143.0	14	30	11	53	(115.0–119.9)
132.6	13	26			
115.3	11	53			

Table 1.5. Too Dense a Stem-and-Leaf Plot of Weights of 17 Scrapers from the Black Site

Weight (g)	Stems	Leaves		
148.7	14	87		
154.5	15	45		
169.5	16	95		
145.1	14	51	17	
157.9	15	79	16	12, 47, 95
137.8	13	78	15	19, 20, 45, 79
151.9	15	19	14	13, 30, 51, 62, 69, 87, 93
146.2	14	62	13	26, 78
164.7	16	47	12	
149.3	14	93	11	53
141.3	14	13		
161.2	16	12		
146.9	14	69		
152.0	15	20		
143.0	14	30		
132.6	13	26		
115.3	11	53		

Table 1.5 shows a still denser stem-and-leaf plot of the same numbers. Stem and leaf sections are separated as in Table 1.4, but only one position is allowed on the stem for each stem section. At this scale, the bunching of numbers is still evident, but what seemed an outlier in Table 1.4 has come so close to the bunch that it no longer seems very different. The characteristics of the batch are less clearly displayed in this stem-and-leaf plot because it crowds the numbers too closely together.

Table 1.6 is yet another stem-and-leaf plot of the same numbers. This one is much too dense. There is simply not enough room on the stem for the leaves to spread out far enough to show the patterning. The outlier from Table 1.4 is no longer apparent (although it is still there—it is just obscured by the inappropriate scale). It is difficult even to evaluate the extent of the bunching of numbers. You can create the next step in the direction of denser stem-and-leaf plots for these numbers yourself. It has a stem consisting only of 1, with all the leaves in one line next to it.

An appropriate scale for a stem-and-leaf plot avoids the two extremes seen in Tables 1.3 and 1.6. The leaves should make one or more branches or bunches of leaves that protrude from the stem. This cannot happen if they are spread out along a stem that is simply too long as in Table 1.3. At the same time, the leaves should be allowed to spread out enough so that outliers can be noticed and two or more bunches, if they occur, can be distinguished from one another. This latter cannot happen if the leaves are crowded together as in

Table 1.6. Much Too Dense a Stem-and-Leaf Plot of Weights of 17 Scrapers from the Black Site

Weight	Stems	Leaves		
148.7	1	487		
154.5	1	545		
169.5	1	695		
145.1	1	451		
157.9	1	579		
137.8	1	378		
151.9	1	519		
146.2	1	462	1	519, 520, 545, 579, 612, 647, 695
164.7	1	647	1	153, 326, 378, 413, 430, 451, 462, 469, 487, 493
149.3	1	493		
141.3	1	413		
161.2	1	612		
146.9	1	469		
152.0	1	520		
143.0	1	430		
132.6	1	326		
115.3	1	153		

Table 1.6. Tables 1.4 and 1.5 show stem-and-leaf plots at scales that are clearer, although Table 1.4 definitely shows the patterns more clearly than Table 1.5.

Different statisticians make stem-and-leaf plots in slightly different ways. There are several approaches to spreading out or compressing the scale. The exact format followed is less important than the principle of showing as clearly as possible the patterns to be observed in the batch of numbers. However the stem sections are divided and positioned, it is important that each stem position correspond to a range of numbers equal to that of every other stem section. It would be a bad idea to structure a stem with positions corresponding to, say, 3.0–3.3, 3.4–3.6, and 3.7–3.9, because the intervals are unequal; that is, a larger range is included between 3.0 and 3.3 than in the other two intervals.

The stem-and-leaf plots in this book have lower numbers at the bottom and higher numbers at the top. This makes it easier to talk about numbers and stem-and-leaf plots in the same terms since lower numbers are lower on the plot and higher numbers are higher on the plot. It is more common for stem-and-leaf plots to be drawn with lower numbers at the top and higher numbers at the bottom. Either way, the stem-and-leaf plot shows the same patterns.

Finally, the stem-and-leaf plots in the tables in this chapter have the leaves on each line in numerical order. This makes no difference in observing the kinds of patterns we have been noting here, but it does make it easier to do some of the things we will do with stem-and-leaf plots in the following chapters. It makes drawing a stem-and-leaf plot a little more time-consuming, but it is well worth the effort, as we shall see.

BACK-TO-BACK STEM-AND-LEAF PLOTS

The stem-and-leaf plot is a fundamental tool not just for exploring a single batch but also for comparing batches. The batch of numbers in Table 1.7 consists of post hole diameters from the Smith site, which we may want to compare to the batch of post hole diameters from the Black site (Table 1.1). These batches can be related since they are measurements of the same variable (diameter of post holes), although two different sets of post holes are involved. Table 1.8 shows a *back-to-back stem-and-leaf plot* in which the leaves representing both batches of numbers are placed on opposite sides of the same stem.

We see the bunch of post holes at diameters of 9 to 12 cm that we saw for the Black site in Table 1.2, as well as the outlier, or unusually large post hole 44.6 cm in diameter. For the Smith site we see a bunch of numbers as well, but this bunch of numbers falls somewhat higher on the stem than the bunch for the Black site. We quickly observe, then, that the post holes at the Smith site are in general of larger diameter than those at the Black site. This general pattern is unmistakable in the stem-and-leaf plot even though the 44.6 cm post hole at the Black site is by far the largest post hole in either site. There is also an outlier among post holes at the Smith site—in this instance a low outlier much smaller than the general run of post holes at the site. If this post hole were at the Black site instead of the Smith site, it would not be nearly so unusual, but at the Smith site it is clearly a misfit.

HISTOGRAMS

The stem-and-leaf plot is an innovation of exploratory data analysis. Although it has certainly appeared in the archaeological literature, there is a traditional way of drawing plots with similar information that is probably

Table 1.7. Diameters of 15 Post Holes at the Smith Site (cm)

20.5	18.3	19.4	18.9
17.2	17.9	16.4	16.8
15.3	18.6	18.8	8.4
15.9	14.3	15.7	

Table 1.8. Back-To-Back Stem-and-Leaf Plot of Post Hole Diameters (in cm) from the Black and Smith Sites (Tables 1.1 and 1.7)

Black site		Smith site
6	44	
	43	
	42	
	41	
	40	
	39	
	38	
	37	
	36	
	35	
	34	
	33	
	32	
	31	
	30	
	29	
	28	
	27	
	26	
	25	
	24	
	23	
	22	
	21	
	20	5
	19	4
	18	3689
	17	29
	16	48
	15	379
2	14	3
	13	
9	12	
8741	11	
85	10	
721	9	
	8	4
6	7	

Table 1.9. Areas of 29 Sites in the
Kiskiminetas River Valley

Site areas (ha)	Stem-and-leaf plot	
12.8	15	3
11.5	14	0
14.0	13	49
1.3	12	388
10.3	11	0257
9.8	10	367
2.3	9	089
15.3	8	27
11.2	7	4
3.4	6	
12.8	5	
13.9	4	5
9.0	3	48
10.6	2	0239
9.9	1	37
13.4		
8.7		
3.8		
11.7		
1.7		
12.3		
11.0		
2.9		
10.7		
7.4		
8.2		
2.0		
2.2		
4.5		

more familiar to more archaeologists. It is the histogram, and it corresponds precisely to the stem-and-leaf plot. The histogram is familiar enough that no detailed explanation of it is needed here. Table 1.9 provides a stem-and-leaf plot of the areas of 29 sites in the Kiskiminetas River valley. Figure 1.1 shows that a histogram of this same batch of numbers is simply a boxed-in stem-and-leaf plot turned on its side with the numbers themselves eliminated as leaves. Most of the same patterns we have noted up to now in stem-and-leaf plots can be observed in histograms as well. In making a histogram, one faces the same choice of scale or interval that we have already discussed for the stem-and-leaf plot, and precisely the same considerations apply. Histograms, on the other hand, have the advantage of being somewhat more elegant and esthetically pleasing as well as being more familiar to archaeologists. Stem-

---

Statpacks

The stem-and-leaf plot is such a simple way to display the numbers in a batch that it can be produced quickly and easily with pencil and paper. When working with pencil and paper, it is necessary only to be careful to line the numbers up vertically so that the patterns are represented accurately. It is also easy to use a typewriter or, better yet, a word processor to produce a stem-and-leaf plot. As when working with pencil and paper, it is important to line the numbers up vertically. This happens automatically with a typewriter or word processor as long as the characters printed or displayed on the computer screen are of equal widths. Type styles in which characters are not all the same width (that is, in which 1, for example, is narrower than 2) don't work for stem-and-leaf plots because the numbers will get out of alignment. The easiest way to make stem-and-leaf plots, of course, is with a statistics computer package, or *statpack* for short. A statpack will perform the entire operation automatically, including choosing an appropriate scale or interval for the stem. Some statpacks still do not include exploratory data analysis (EDA) tools like stem-and-leaf plots, but many do.

Histograms are more time-consuming to draw than stem-and-leaf plots, but many computer programs will draw them very presentably. True statistical packages are best for this task, since their programmers had in mind exactly the goals discussed in this chapter when they wrote the programs. The numerous programs that specialize in drawing multicolored or three-dimensional bar charts might at first glance seem a more attractive way to accomplish these aims. Such charts, while much more striking visually, and therefore very popular in newspapers and magazines, are not exactly histograms and can actually be more difficult to read precisely. They do not lend themselves very well to exploration for patterns in a batch of numbers or to comparison of several batches.

---

and-leaf plots, on the other hand, have the advantage that the full detail of the actual numbers is present, and this makes it possible to use them in ways that histograms cannot be used, as we shall see in the next few chapters. In general terms, however, the stem-and-leaf plot and the histogram serve fundamentally the same purpose.

MULTIPLE BUNCHES OR PEAKS

The batch of numbers in Table 1.9 also demonstrates another characteristic of batches that sometimes becomes obvious in either a stem-and-leaf plot or a histogram. We see the usual bunching of numbers in the stem-and-

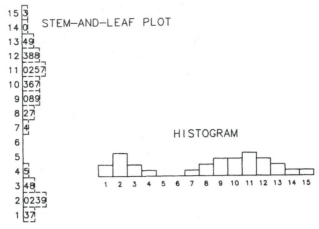

```
15 |3
14 |0      STEM—AND—LEAF  PLOT
13 |49
12 |388
11 |0257
10 |367
 9 |089
 8 |27
 7 |4                        HISTOGRAM
 6 |
 5 |
 4 |5
 3 |48
 2 |0239
 1 |37
```

Figure 1.1. A histogram of areas of 29 sites in the Kiskiminetas River valley.

leaf plot. In this case, however, there are two distinct and separate bunches, one between about 1 and 4 ha and another between about 7 and 15 ha. The same bunches are obvious in the histogram (Figure 1.1), where the two separate bunches appear as two hills or *peaks*. Such a pattern of multiple bunches or peaks is a clear indication of distinct kinds of cases—in this instance two distinct kinds of sites. We might likely call them large sites and small sites, and the pattern seen in the stem-and-leaf plot or the histogram indicates that the two are clearly separate. That is, in discussing these as large and small sites, we would not be arbitrarily dividing sites up into large and small but rather responding to an innate characteristic of this batch of numbers. We see quickly that the large sites are more numerous, but there are enough small sites to form a clear and separate peak. This is not a case of outliers but, instead, of two sets of sites, each numerous enough to form its own peak in the histogram.

The presence of multiple peaks in a batch is always an indication that two or more fundamentally different kinds of things have been thrown together and measured. To take a ridiculous example, I might measure the diameters of a series of dinner plates and manhole covers. If I presented these as a single list of measurements of round objects, you would see immediately in a stem-and-leaf plot that there were two separate peaks. Knowing nothing about the objects except their diameters, you would guess that two fundamentally different kinds of things had been measured. You would be correct to subdivide the batch into two batches with no further justification than the pattern you saw in the stem-and-leaf plot. One of the first things you might do, however, would be to seek further information about the nature of the objects that might clarify their differences. Your reaction, on finding out that both dinner plates and manhole covers were included among the objects measured, might well be, "No wonder; now I understand!" This is a perfectly appropriate reaction and would put substance behind a division made on purely formal grounds (that is, on the basis of the pattern observed in a stem-and-leaf plot).

To repeat, batches with multiple peaks cannot be analyzed further. The only correction for this problem is to subdivide the batch into separate batches for separate analysis. In the best of all possible worlds, we can identify other characteristics of the objects in question to aid us in making the division. If not, we must do it simply on the basis of the stem-and-leaf plot or histogram, drawing a dividing line on the number scale at the lowest point of the valley that separates the peaks. This is especially easy for the numbers illustrated in Figure 1.1. The lowest point of the valley here is at 5 and 6 ha. There are no sites at all of this size, so the small sites are clearly those ranging from 1 to 5 ha, and the large sites are those ranging from 7 to 16 ha. If there is not an actual gap at the bottom of the valley, as there is in this instance, just where to draw the dividing line may not be so obvious, but it must be done nevertheless before proceeding to any further analysis.

PRACTICE

In Tables 1.10 and 1.11 are two batches of numbers—measurements of the lengths of scrapers recovered from two sites. The scrapers are made from either flint or chert. These numbers could be considered a single batch of numbers (lengths of scrapers, disregarding what raw material they were made from and what site they came from). They also form two related batches in two different ways. We could divide the single batch into two batches according to the site at which the scrapers were recovered. (This is

Table 1.10. Scrapers from Pine Ridge Cave

Raw material	Length (mm)	Raw material	Length (mm)
Chert	25.8	Chert	25.9
Chert	6.3	Chert	23.8
Flint	44.6	Chert	22.0
Chert	21.3	Chert	10.6
Flint	25.7	Flint	33.2
Chert	20.6	Chert	16.8
Chert	22.2	Chert	21.8
Chert	10.5	Flint	48.3
Chert	18.9		

Table 1.11. Scrapers from the Willow Flats Site

Raw material	Length (mm)	Raw material	Length (mm)
Chert	15.8	Flint	49.1
Flint	39.4	Flint	41.7
Flint	43.5	Chert	15.2
Flint	39.8	Chert	21.2
Chert	16.3	Flint	30.2
Flint	40.5	Flint	40.0
Flint	91.7	Chert	20.2
Chert	21.7	Flint	31.9
Chert	17.9	Flint	42.3
Flint	29.3	Flint	47.2
Flint	39.1	Flint	50.5
Flint	42.5	Chert	10.6
Flint	49.6	Chert	23.1
Chert	13.7	Flint	44.1
Chert	19.1	Flint	45.8
Flint	40.6		

the way the numbers are presented in the tables.) Or we could divide the single batch into two batches according to the raw material they were made of (disregarding which site they came from).

1. Make a stem-and-leaf plot of scraper lengths, treating the entire set of scrapers as a single batch. Experiment with different intervals for the stem to consider which interval produces the most useful plot. What patterns do you see in the plot?
2. Make a back-to-back stem-and-leaf plot of scraper lengths, treating the scrapers from the Willow Flats site as one batch and those from Pine Ridge Cave as another batch. (That is, ignore the raw material of which the scrapers were made for the moment.) How do the two batches compare to each other? Do you see any patterns that help you interpret the stem-and-leaf plot of all scrapers as a single batch?
3. Make a back-to-back stem-and-leaf plot of scraper lengths, treating the flint scrapers as one batch and the chert scrapers as another batch. (That is, this time ignore which site the scrapers came from.) How do these two batches compare to each other? Do you see any patterns this time that help you interpret the stem-and-leaf plot of all scrapers as a single batch?

Chapter *2*

The Level, or Center, of a Batch

As we saw in Chapter 1, the numbers in a batch often bunch together. If we compare two related batches of numbers, the principal bunch in one batch may well have higher numbers in general than the principal bunch in the other batch. We can say that the batches have different *levels*, or *centers*. It is convenient to use a numerical index of the level for such comparisons. The several such indexes in common use are traditionally referred to as *measures of central tendency*.

THE MEAN

The most familiar index of the center of a batch is the *mean*, outside statistics more commonly referred to as the *average*. Calculation of the mean is just as we all learned in elementary school: the sum of all the numbers in the batch is divided by the number of numbers in the batch. Since this is such a familiar calculation, it provides a good opportunity to introduce some mathematical notation that is particularly useful in statistics. The equation expressing the calculation of the mean is

$$\bar{X} = \frac{\Sigma x}{n}$$

17

where

x represents each number in a batch, individually;
n is the number of x's;
$\bar{X}$ is the mean or average of x (pronounced "x bar").

The Greek letter Σ (capital sigma) stands for "the sum of," and is a symbol used frequently in statistics (Σx simply means "the sum of all the x's"). Formulas with Σ may seem formidable, but, as we have just seen, Σ is simply shorthand for a relatively simple and familiar calculation. Sigma is virtually the only mathematical symbol used in this book that is not common in basic algebra.

Table 2.1 presents some data on weights of flakes recovered from two bell-shaped storage pits in the same site. The back-to-back stem-and-leaf plot reveals that the flakes from Pit 1 bunch together between about 9 and 12 g, with one outlier at 28.6 g (to which we probably do not want to pay too much attention). The flakes from Pit 2 also bunch together, although the peak is more spread out and may even have some tendency to split in two. The center of the batch of flakes from Pit 2 would appear to be a little higher

Table 2.1. Weights of Flakes Recovered from Two Bell-Shaped Pits

	Flake weights (g)		Back-to-back stem-and-leaf plot		
	Pit 1	Pit 2	Pit 1		Pit 2
	9.2	11.3	6	28	
	12.9	9.8		27	
	11.4	14.1		26	
	9.1	13.5		25	
	28.6	9.7		24	
	10.5	12.0		23	
	11.7	7.8		22	
	10.1	10.6		21	
	7.6	11.5		20	
	11.8	14.3		19	
	14.2	13.6		18	
	10.8	9.3		17	
		10.9		16	
				15	
$\bar{X}$	12.33	11.42	2	14	13
Md	11.10	11.30		13	56
			9	12	0
		$\bar{X}$ —— 874		11	35 —— $\bar{X}$
		Md —— 851		10	69 —— Md
			21	9	378
				8	
			6	7	8

on the whole than for those from Pit 1. For the flakes from Pit 1, the mean (calculated by summing up all 12 weights and dividing the total by 12) is 12.33 g. For Pit 2, the mean (calculated by summing up all 13 weights and dividing the total by 13) is 11.42 g. Both means are indicated in their approximate positions along the stem in the stem-and-leaf plot.

We can be fairly happy with the mean as an index of the center for Pit 2; it does point to something like the center of the main bunch in the batch, as seen in the stem-and-leaf plot. When we look at Pit 1, however, we have cause for concern. The mean seems to be well above the center of the main bunch in the batch. It is "pulled up" quite strongly by the high outlier at 28.6 g, which has a major impact on the sum of the weights. Since we just observed that the Pit 1 batch has a somewhat lower level than the Pit 2 batch, it is alarming that the mean for Pit 1 is actually higher than the mean for Pit 2. A comparison of means for these two batches would suggest that flakes from Pit 1 tended to weigh more than those from Pit 2—a conclusion exactly opposite to the one we arrived at by examining the stem-and-leaf plot. In this instance, the mean is not behaving very nicely. That is, it is not providing a useful index of the center of the Pit 1 batch for the purpose of comparing that batch to the Pit 2 batch.

THE MEDIAN

The *median* is sometimes a more useful index of the center of a batch. The median is simply the middle number in the batch (if the batch contains an odd number of numbers) or halfway between the two middle numbers (if it contains an even number of numbers). The stem-and-leaf plot is particularly useful for finding the median, because it enables us to count in from either the top or the bottom to the middle number. It is especially important to place the leaves in numerical order on each line of the stem-and-leaf plot in order to do this without errors. The alternative to the stem-and-leaf plot, the histogram, cannot be used for finding the median because the histogram does not contain the actual numbers.

To find the median weight of flakes from Pit 1, we first count the number of flakes. Since there are 12 (an even number), the median will be halfway between the middle two numbers. The middle two numbers will be the sixth and seventh, counting in from either the highest or lowest number. For example, counting leaves in the stem-and-leaf plot for Pit 1 from the bottom, or lowest number, we have the first five numbers: 7.6, 9.1, 9.2, 10.1, and 10.5; then the sixth and seventh numbers: 10.8 and 11.4. Alternatively, counting leaves from the top, or highest number, we have the first five numbers: 28.6, 14.2, 12.9, 11.8, and 11.7; then the sixth and seventh: 11.4 and 10.8, the same as before. Halfway between 10.8 and 11.4 is 11.1. So the median weight of flakes from Pit 1 is 11.10 g ($Md = 11.10$ g).

For Pit 2, there are 13 flakes, so the median will be the middle number, or the seventh in from either the highest or lowest. Counting leaves from the top gives us the first six numbers: 14.3, 14.1, 13.6, 13.5, 12.0, and 11.5; then the seventh: 11.3. Counting leaves from the bottom gives us the first six numbers: 7.8, 9.3, 9.7, 9.8, 10.6, 10.9; then the seventh: 11.3, exactly as before. Thus the median weight of flakes from Pit 2 is 11.30 g ($Md = 11.30$ g).

Medians for both batches are indicated on the stem-and-leaf plot, and both indicate points that are visually more satisfying indications of the centers of the two batches. Comparing the levels of the two batches according to their medians also seems more reasonable than our attempt to use their means for this purpose. The median weight of flakes in Pit 2 is slightly higher than that for Pit 1, which is indeed the conclusion we came to based on observation of the general pattern of the stem-and-leaf plot.

OUTLIERS AND RESISTANCE

It might seem surprising that the mean and the median behave so differently in this example. After all, both are fairly widely used indexes of the level of a batch. And yet, comparing the two batches in this example by means and by medians gave opposite conclusions about which batch had a higher center. Clearly, it is the mean of the flakes from Pit 1 that seems strange. Its peculiarly high position is attributable entirely to the effect that the one high outlier (the flake that weighs 28.6 g) has on the calculations. While it pulls the mean up substantially, this outlier, in contrast, has no effect whatever on the median. If instead of weighing 28.6 g, this flake had weighed 12.5 g, the median flake weight for Pit 1 would not have changed at all. (If this sounds strange to you, try it yourself with pencil and paper. Make a new stem-and-leaf for the Pit 1 batch including 12.5 g instead of 28.6 g in the list. You will soon appreciate that it does not matter at all how you change the numbers in either the upper half or the lower half of the batch. As long as you do not make a change that moves a number from the upper half to the lower half or vice versa, the median remains exactly the same.)

This is one example of a general principle. The mean of a batch is always strongly affected by any outliers that may be present. The median is entirely unaffected by them. In statistical jargon, the median is very *resistant*. The mean is not at all resistant.

ELIMINATING OUTLIERS

Outliers, then, present a serious problem to using the mean as an index of the center of a batch. It would be nice to eliminate outliers if we could, and, it turns out, often we can. We should always pay special attention to outliers. Often they indicate errors in data collection or recording. This possibility was already broached in Chapter 1. It was suggested that the extraordinarily large post hole in the example in Table 1.2 might have been the result of an error in measurement or in data recording. Such an error could be corrected by reference to photographs and drawings of the excavation, thus eliminating the outlier.

Even if it turns out that an outlier is, indeed, a correct value, it still may be desirable to eliminate it. A classic example of such a situation concerns the mail-order clothing firm of L.L. Pea, Inc., specializing (of course) in the famous Pea coat. L.L. Pea employs ten shipping clerks, nine of whom are each paid $8.00 per hour while the tenth earns $52.00 per hour. The median wage in the L.L. Pea shipping room, then, is $8.00 per hour, while the mean wage is $12.40 per hour. Once again, the mean has been raised substantially by an outlier, while the median has been entirely unaffected. A careful check of payroll records reveals that it is, indeed, true that nine shipping clerks are paid $8.00 per hour while one earns $52.00 per hour. It also reveals, however, that the highly paid clerk is Edelbert Pea, nephew of L.L., the founder of the company, who spends most of his "working" hours in the company cafeteria anyway. If our interest is in the wages of shipping clerks, there is clearly no reason to include young Edelbert among our data. We are much better off simply to eliminate him as not truly a case of what we wish to study and deal as may be appropriate with the data on the other nine shipping clerks.

It is often sensible to eliminate outliers in just such a manner. If a good reason can be found aside from just the aberrant number in the data (as in the instance of Edelbert Pea), we can feel quite comfortable about eliminating outliers. In the example batch in Table 2.1 for Pit 1, perhaps we would note that the unusually heavy flake was of a very different form from all the rest or of a very different raw material. In this last case, we might reduce our batch to obsidian flakes, say, rather than all flakes, in order to eliminate a single very heavy chert flake. Even if such external reasons cannot be found to justify it, a distant outlier can be eliminated simply on the basis of its measurement. There are, however, other treatments that take care of outliers without making it seem that somehow we are fudging our data by leaving out cases we don't like.

THE TRIMMED MEAN

The *trimmed mean* systematically removes extreme values from both upper and lower ends of a batch in a balanced fashion. In considering the level of a batch, it is the central bunch of numbers that matters most. It is not uncommon for the highest and lowest numbers to straggle away from this bunch in an erratic manner, and it is important not to be confused by such unruly behavior on the part of a few numbers. The trimmed mean effectively avoids such confusion by simply eliminating some proportion of the highest and lowest numbers in the batch from consideration.

For example, we might calculate a 5% trimmed mean of the flake weights from Pit 1 in Table 2.1. For a 5% trimmed mean, we eliminate the highest 5% of the batch and the lowest 5% of the batch. There are 12 numbers in this batch, so we remove 5% of 12 numbers from each end. Since $.05 \times 12 = .60$, and .60 rounds up to 1, we remove one number from the top and one number from the bottom. That is, we remove the highest number (28.6) and the lowest number (7.6) from the batch. (In deciding how many numbers to remove for the trimmed mean we always round *up*.) After removing the highest and lowest numbers, we have a trimmed batch of 10 numbers ($n_T = 10$). The trimmed mean is simply the ordinary mean of the remaining 10 numbers, once the highest and lowest have been removed. For Pit 1 the 5% trimmed mean, $\bar{X}_T$, is the sum of the remaining numbers divided by n_T (that is, 10), or 11.17 g. For Pit 2, a 5% trimmed mean also requires eliminating a single number from each end of the batch ($.05 \times 13 = .65$, which rounds up to 1). The total of the remaining numbers is divided by n_T (that is, 11), for $\bar{X}_T = 11.48$ g.

We can see that the trimmed mean, unlike the ordinary mean, is resistant to the effect of outliers. In this example, the 5% trimmed means are quite similar to the medians. They would lead us to conclude that flakes in Pit 2, in general, weigh slightly more than flakes in Pit 1, just as observation of the stem-and-leaf plot makes us know we should conclude.

In the 5% trimmed mean calculated above, 5% is the *trimming fraction*. The trimming fraction can be adjusted to fit the needs of a particular situation. Customarily, the trimming fraction is some multiple of 5% (5%, 10%, 15%, etc.). The most frequently used trimming fractions are probably 5% and 25%. The 25% trimmed mean is sometimes called the *midmean* because it is the mean of the middle half of the numbers (one-fourth of the numbers having been eliminated from the top of the batch and one-fourth from the bottom).

As one final example, a 25% trimmed mean of the flake weights from Pit 1 in Table 2.1 requires elimination of the three highest and the three lowest numbers ($.25 \times 12 = 3$). The mean of the remaining six numbers is 11.05 g. For the flake weights from Pit 2, a 25% trimmed mean requires removal of four numbers from the top and bottom ($.25 \times 13 = 3.25$, which rounds up to 4). The mean of the remaining five numbers is 11.26 g. Just as with the 5% trimmed mean, the undesirable effects of outliers have been avoided entirely, and the comparison of means shows that Pit 2 flakes are, in general, slightly heavier than Pit 1 flakes.

Statpacks

Any statistics package will determine the mean and median for a batch of numbers. Not very many, however, provide the trimmed mean as a defined option. What you may have to do to get your statpack to calculate a trimmed mean is do the trimming yourself. You could simply omit the numbers to be trimmed when entering the data initially or you could delete those cases (or code them as missing data by whatever provision your statpack makes for handling missing data). Then your statpack can easily calculate the mean of the remaining numbers.

It is worth noting that the median could be thought of as the ultimate in trimmed means, the 50% trimmed mean. Removing the upper half of the batch and the lower half of the batch leaves nothing but the midpoint, or median.

WHICH INDEX TO USE

The median, the mean, and the trimmed mean are all numerical indexes of the center of a batch. The question thus arises, which one should we use? This question has no simple answer. Sometimes it is better to use the mean, sometimes the median, sometimes the trimmed mean. It depends on the characteristics of the batch in question and on what you intend to do with the numerical index of the center once you have it. The mean is the most familiar, and that is an advantage worth considering, since just about anyone feels comfortable if you tell them what the mean of a batch of numbers is. If the batch does not have outliers that make the mean a deceptive value, then it may well be the best choice. The median is slightly less familiar, but it is highly resistant, and so it is used fairly often for batches with outliers. The trimmed mean is considerably less familiar to most archaeologists, but it combines advantages of mean and median in some respects.

As we will see in later chapters, the mean has some special properties that make it highly useful in statistics. It is thus often tempting to use the mean, even when the batch has outliers that affect it. The trimmed mean can be put to work in at least some of the same ways the mean can, however, without interference from outliers. That is what makes the trimmed mean worth discussing, even though it is more complicated to calculate than either the mean or the median and less well known among archaeologists. The median, unfortunately, cannot be used in any of these special ways. Even though it is quite straightforward and useful for the initial task of comparing batches, then, the median will not be as important to us farther along in this book as the mean and the trimmed mean.

BATCHES WITH TWO CENTERS

Sometimes examination of a stem-and-leaf plot makes it clear that a batch contains two or more quite distinct bunches, as discussed in Chapter 1. We will call such batches *two-peaked* or *multipeaked*. (The metaphor of the peak is derived from the histogram, where a bunch of numbers resembles a hill or peak, but it is easy enough to think of a stem-and-leaf plot in these terms as well.)

Table 2.2 provides the areas (in square meters) of structures excavated at the Black–Smith sites. The stem-and-leaf plot shows that these structures form two separate groups on the basis of their areas. There are large structures, mostly from about 15 to 21 m^2, and small structures, from about 3 to 7 m^2. It would make little sense to talk about the center of this batch, because it clearly has two centers. If it makes little sense to talk about its center, then it makes even less sense to calculate a numerical index of its center. If we tried it, the results would be nonsense. The mean, for example, of the batch in Table 2.2 is 12.95 m^2. This value falls in between the two distinct groups, characterizing no structures at all. At 15.15 m^2, the median also fails to characterize the center of anything meaningful.

In Chapter 1 it was asserted that a batch with two peaks cannot be analyzed as a single batch but must be separated into two batches for separate analysis. Thinking about indexes of center makes it clear why. The rule that batches with multiple peaks must be separated for analysis is not a rule that must be memorized. It is simply the only practice that makes sense to any analyst who keeps firmly in mind what indexes of center are doing and how they behave. In a case like this, one must think that there are basically two different kinds of structures represented, perhaps houses and grain bins. Other information concerning these structures could be examined for evidence relevant to such a notion. In any event, before further quantitative analysis the batch must be broken into two batches and the large structures treated separately from the small structures. We would make the break at about 10 or 11 m^2 in the middle of the large gap visible in the stem-and-leaf plot. The 16 small structures, then, that are less than 10 m^2, have a mean area of 5.67 m^2 (and an almost identical median area of 5.70 m^2). The 20 large structures have a mean area of 18.77 m^2 (and, once again, an almost identical median area of 18.75 m^2). For both small structure areas and large structure areas, then, both the mean and the median provide meaningful and useful indexes of the center. (Locate them along the stem in the stem-and-leaf plot, and you will see that they are indeed in the center of the main bunch of numbers for each subbatch.) Breaking a two-peaked batch into two batches has made it possible to calculate numerical indexes of the centers of the two batches that are sensible—something that we could not do with the original batch taken as a whole.

Table 2.2. Floor Areas of Structures at the Black–Smith Sites

Areas (m^2)	Stem-and-leaf plot	
18.3	26	8
18.8	25	
16.7	24	
6.1	23	4
5.2	22	
21.2	21	2
19.8	20	07
4.2	19	128
18.3	18	33789
3.6	17	59
20.0	16	27
7.5	15	03
15.3	14	
26.8	13	6
5.4	12	
18.7	11	
6.2	10	
7.0	9	
20.7	8	
18.9	7	05
19.2	6	1277
6.7	5	244689
19.1	4	259
23.4	3	6
4.5		
16.2		
5.6		
17.5		
5.9		
6.7		
4.9		
17.9		
15.0		
13.6		
5.4		
5.8		

Batches like the one in Table 2.2 are often referred to loosely as *bimo-dal*, after the term *mode*, which refers to the single most common category in a stem-and-leaf plot or histogram. Sometimes the mode is used as an index of the center of a batch. In Table 2.2, the mode would be at about 5 m², where six structures fall. This, clearly, is something like the center of the batch of small structures, but it won't do as an index of the center of the entire batch. There is a secondary mode at about 18 m², where five structures fall. This is something like the center of the batch of large structures. Only if exactly the same number of structures fell at 5 m² and at 18 m² would this batch really have two modes. Strictly speaking, it has a mode and a secondary mode rather than two modes. Nevertheless, such multipeaked batches are often referred to as bimodal.

PRACTICE

1. Look back at the data on scraper lengths given in Tables 1.10 and 1.11. Calculate indexes of the centers of these two batches of measurements (that is, for Pine Ridge Cave and for the Willow Flats site). Try out the mean, the median, and a trimmed mean (with whatever trimming fraction you think is most appropriate). Which index of center makes most sense for comparing scraper lengths from these two sites? Why? (Note that if your aim is to compare the two sites in regard to the lengths of scrapers found there, you must use the same index for both. You can't compare the mean length for one site to the median length for the other.) Summarize the comparison of scraper lengths you have made between the two sites. That is, what has all this told you about scraper lengths at the two sites?

2. Using the data in Tables 1.10 and 1.11 once again, calculate indexes of the centers for flint scrapers and for chert scrapers, disregarding which site the scrapers come from. Try the mean, the median, and the trimmed mean again. Which index makes most sense for comparing the lengths of scrapers made of different raw materials? Why? How would you summarize all together the comparisons you have made between flint and chert scrapers and between the Willow Flats site and Pine Ridge Cave?

Chapter *3*

The Spread, or Dispersion, of a Batch

Some batches of numbers are very tightly bunched together while others are much more spread out. This property is referred to in exploratory data analysis as *spread* (or in more traditional statistical terms as *dispersion*), and it is often an informative characteristic of a batch to which you should pay attention. Just as it is convenient to have a numerical index for the level or center of a batch, it is also convenient to have a numerical index for the spread, or dispersion, of a batch. Once again there are several different numerical indexes that behave differently and are thus used in different circumstances.

THE RANGE

The simplest index of the spread of a batch is its *range*. The range in statistics is exactly what it is in everyday conversation: the difference between the lowest number and the highest number in the batch. Table 3.1 presents the same example numbers we discussed in the previous chapter. The range for the weights of flakes recovered from Pit 1 is the difference between 28.6 g and 7.6 g, or 21.0 g (28.6 g – 7.6 g = 21.0 g). The range for the weights of flakes recovered from Pit 2 is the difference between 14.3 g and 7.8 g, or 6.5 g (14.3 g – 7.8 g = 6.5 g).

Table 3.1. Weights of Flakes Recovered from Two Bell-Shaped Pits

	Flake weights (g)		Back-to-back stem-and-leaf plot		
	Pit 1	Pit 2	Pit 1		Pit 2
	9.2	11.3	6	28	
	12.9	9.8		27	
	11.4	14.1		26	
	9.1	13.5		25	
	28.6	9.7		24	
	10.5	12.0		23	
	11.7	7.8		22	
	10.1	10.6		21	
	7.6	11.5		20	
	11.8	14.3		19	
	14.2	13.6		18	
	10.8	9.3		17	
		10.9		16	
				15	
$\bar{X}$	12.33	11.42	2	14	13
Md	11.10	11.30		13	56
			9	12	0
range	21.0	6.5	874	11	35
midspread	3.7	3.7	851	10	69
			21	9	378
				8	
			6	7	8

We notice immediately that the range suffers from the same problem that the mean suffers from: it is not at all resistant. In fact, it is even less resistant than the mean. Not only is it strongly affected by outliers, but it is also likely to depend entirely on outliers. Examination of the stem-and-leaf diagram reveals how misleading the range is in this instance. The two batches here have rather similar spreads, but we would probably say that the flake weights from Pit 2 are more spread out than those of Pit 1 because the central bunch (which is always the most important part of the batch) is more dispersed along the stem. Nevertheless, the range for Pit 1 is much greater, entirely because of the one very high outlier in the Pit 1 batch. Although the range is simple to calculate and easily understood by everyone, it is likely to be very misleading unless all outliers can be removed.

THE MIDSPREAD, OR INTERQUARTILE RANGE

The *midspread* is the range of the middle half of a batch. The highest 25% of the numbers and the lowest 25% of the numbers are thus disregarded.

It could be thought of as a sort of trimmed range, thinking back to the trimmed mean discussed in Chapter 2.

In practice the midspread is found by locating the *quartiles* and subtracting the lower quartile from the upper quartile. The *upper quartile* is something like the median of the upper half of the batch and the *lower quartile* is something like the median of the lower half of the batch, although the rules used for finding the quartiles differ slightly from those used for finding the median. (In exploratory data analysis the quartiles are often called the *hinges*.) To find the quartiles, first divide the number of numbers in the batch by 4. If the result is a fraction, round it up to the next whole number. Then count in that many numbers from the highest and lowest number in the batch.

For example, there are 12 flakes from Pit 1 for which weights are given in Table 3.1. We divide 12 by 4 and get 3. The upper quartile is the third number from the top of the stem-and-leaf, or 12.9 g. The lower quartile is the third number from the bottom of the stem-and-leaf, or 9.2. The midspread is then 12.9 g − 9.2 g = 3.7 g. For Pit 2, we have a batch of 13 weights; (13/4) = 3.25, which we round up to 4. The upper quartile is the fourth number from the top of the stem-and-leaf, or 13.5 g. The lower quartile is the fourth number from the bottom of the stem-and-leaf, or 9.8 g. The midspread is thus 13.5 g − 9.8 g = 3.7 g.

The midspread gives us better results for this example than the range, indicating that both batches are spread out to the same degree (a midspread of 3.7 g for both batches). This is at least closer to the mark than using a numerical index that shows the Pit 1 batch to be much more spread out than the Pit 2 batch.

The procedure for finding the midspread also reveals why it is sometimes called the *interquartile range* (at least by those who never use two syllables when five will do). The midspread is simply the range between the quartiles, and interquartile range is the traditional term for it. The midspread is used more in exploratory data analysis than in traditional statistics, and it works particularly well with the median to give us a quick indication of the level and spread of a batch.

THE VARIANCE AND STANDARD DEVIATION

The *variance* and *standard deviation* are based on the mean. They are considerably more cumbersome to calculate than the range or the midspread, and they lack some of the immediately intuitive meaning that the range and midspread have. They have technical properties, however, that make them extraordinarily useful, and so they will be of considerable importance to many of the following chapters.

The basic concept on which the variance is based is that of difference from the mean. Clearly the vast majority of numbers in a batch are likely to be rather different from the mean of the batch. We can easily see how different any number in a batch is from the mean by subtracting the mean from it. The first two columns of Table 3.2 illustrate this procedure for all the numbers in the batch of weights of flakes from Pit 2 in Table 3.1. As is logical, the higher numbers in the batch have positive deviations from the mean (because they are *above* the mean), and the lower numbers have negative deviations from the mean (because they are *below* the mean). The numbers at the extreme ends of the batch, of course, deviate quite strongly from the mean in either a positive or negative direction. The more spread out a batch is, the more strong deviations from the mean there are.

If we want to summarize these deviations numerically, it might occur to us to take the mean of the deviations. This won't do, however, because we can see that the deviations must always add up to 0; hence their mean will always be 0. Indeed, a different way to think of the mean is to consider it a "balance point" that makes these deviations add up to 0. (You may notice that the second column of Table 3.2 actually adds up to –0.06 rather than 0. This is a consequence of rounding error, which commonly occurs. All of the deviations are rounded off to two digits following the decimal point, and in this case by pure chance a little more rounding down has occurred than rounding up.)

What we are interested in, as an index of spread, is the set of deviations from the mean without their signs. We could simply drop the signs and add up the absolute values of the deviations, but it turns out to be preferable to get rid of the signs by squaring the deviations from the mean. (The squares of the deviations from the mean are, of course, all positive, as squares must all be.) This calculation is shown in the third column of Table 3.2. It is this third column that we sum up. This sum is sometimes referred to as the sum of the squared deviations from the mean, or simply the *sum of squares*.

This sum of squares will, other things being equal, be larger for a larger batch of numbers than for a smaller batch because a larger batch has more deviations to add up. To arrive at an index that is not affected by the size of the batch but only by its spread, what we need is something like the average squared deviation from the mean. Instead of dividing the sum of squares by the number of numbers in the batch, however, we divide it by one less than the number of numbers in the batch. We do this for purely technical reasons to make the result more useful in future chapters where we take batches of numbers to be samples from larger populations. The equation for the variance, then, is

$$s^2 = \frac{\Sigma(x - \bar{X})^2}{n - 1}$$

Table 3.2. Calculating the Standard Deviation of Flake Weights from Pit 2 (Table 3.1)

x	Deviations from mean $x - \bar{X}$	Squared deviations from mean $(x - \bar{X})^2$
14.3	2.88	8.29
14.1	2.68	7.18
13.6	2.18	4.75
13.5	2.08	4.33
12.0	0.58	0.34
11.5	0.08	0.01
11.3	−0.12	0.01
10.9	−0.52	0.27
10.6	−0.82	0.67
9.8	−1.62	2.62
9.7	−1.72	2.96
9.3	−2.12	4.49
7.8	−3.62	13.10
$\bar{X} = 11.42$	$\Sigma(x - \bar{X}) = -0.06$	$\Sigma(x - \bar{X})^2 = 49.02$ (sum of squares)

$$s^2 = \frac{\Sigma(x - \bar{X})^2}{n - 1} = \frac{49.02}{12} = 4.09$$

$$s = \sqrt{s^2} = \sqrt{4.09} = 2.02$$

where

s^2 is the variance of x;
$\bar{X}$ is the mean of x;
n is the number in the batch of x.

Table 3.2 provides an example of the calculations that correspond to this equation.

The variance has a rather arbitrary character compared to the range or the midspread. The value of the variance is not as easy to relate intuitively to the values in the batch as was the case with the range or midspread. We can at least remove the confusing effect of squaring the deviations by taking the square root of the variance. The result is s, the standard deviation:

$$s = \sqrt{s^2} = \sqrt{\frac{\Sigma(x - \bar{X})^2}{n - 1}}$$

The standard deviation, unlike the variance, is at least expressed in the same units as the original batch. Thus it is appropriate to think of the standard deviation of the weights of flakes from Pit 2 as not just 2.02, but 2.02 g. If we relate the standard deviation to the stem-and-leaf plot in Table 3.1, we see that the standard deviation delineates the portion of the stem within which most of the flake weights fall. That is, most weights are within 2.02 g above or below the mean of 11.42 g, which is to say, most of the weights are between 9.40 g (11.42 g − 2.02 g = 9.40 g) and 13.44 g (11.42 g + 2.02 g = 13.44 g). These two numbers (9.40 g and 13.44 g) provide an approximation of the limits of the main bunch of numbers. That is what it means to say that most of the flake weights are within one standard deviation of the mean. Only a few fall farther than one standard deviation from the mean, that is, farther than 2.02 g from the mean. We can (and will) specify much more about this way of using the standard deviation in later chapters. For the moment, suffice it to say that the standard deviation often provides just this kind of indication about the spread of a batch.

The standard deviation does not behave so satisfactorily for the flake weights from Pit 1. Table 3.3 shows the calculation of the standard deviation for this batch. When we first compared these two batches of numbers (the weights of flakes from Pits 1 and 2) on the basis of the stem-and-leaf plots in Table 2.1, we noted that the flake weights from Pit 1 were (except for the high outlier) more closely bunched up than those from Pit 2. The variance and standard deviation for flake weights from Pit 1, however, are much larger than those for Pit 2, indicating a much larger spread for the flakes from Pit 1—exactly opposite the conclusion the stem-and-leaf plot clearly indicates.

Table 3.3 shows very clearly why the variance and standard deviation are so large for Pit 1: the value for the one heaviest flake deviates very strongly from the mean. That one flake is alone responsible for such a high sum of squares and thus for such a high variance and standard deviation. Clearly, like the mean, the variance and standard deviation are not at all resistant to the effects of outliers. Using the variance or the standard deviation as a numerical index of the spread of a batch, then, is not a good idea at all if the batch has outliers.

Table 3.3 also provides a convenient illustration of why the mean lacks resistance that amplifies the observations made in Chapter 2. Thinking of the mean as a balance point that will make the deviations from the mean sum up to 0 and looking at the second column in Table 3.3 reveals clearly the impact that the one outlier has on the mean as well. In order to balance the effect of that single strong positive deviation the mean has moved far up the number scale, causing the vast majority of the numbers in the batch to be below the mean. It was precisely this undesirable effect that we complained about in Chapter 2.

Table 3.3. Calculating the Standard Deviation of Flake
Weights from Pit 1 (Table 3.1)

x (g)	Deviations from mean $x - \bar{X}$	Squared deviations from mean $(x - \bar{X})^2$
28.6	16.27	264.71
14.2	1.87	3.50
12.9	0.57	0.32
11.8	−0.53	0.28
11.7	−0.63	0.40
11.4	−0.93	0.86
10.8	−1.53	2.34
10.5	−1.83	3.35
10.1	−2.23	4.97
9.2	−3.13	9.80
9.1	−3.23	10.43
7.6	−4.73	22.37
$\bar{X}$ = 12.33	$\Sigma(x - \bar{X}) = -0.06$	$\Sigma(x - \bar{X})^2 = 323.33$ (sum of squares)

$$s^2 = \frac{\Sigma(x - \bar{X})^2}{n - 1} = \frac{323.33}{11} = 29.39$$

$$s = \sqrt{s^2} = \sqrt{29.39} = 5.42$$

THE TRIMMED STANDARD DEVIATION

The basic idea of the trimmed standard deviation is exactly like that of the trimmed mean: outliers are excluded from the sample so that they will not have an undue effect on the result. Calculation of the trimmed standard deviation, however, becomes more involved. Instead of simply reducing the size of the batch by trimming off numbers at the top and bottom, we must maintain the size of the batch by replacing trimmed numbers with the numbers next in line for trimming. Table 3.4 shows this process for calculating a 5% trimmed standard deviation of the batch of flake weights from Pit 1. When, in Chapter 2, we calculated the 5% trimmed mean of this same batch, we trimmed the single highest and lowest number from the batch. This time, we replace the highest number with the next highest number (the highest number that remained in the batch after trimming). Thus 28.6 g becomes 14.2 g. Similarly, we replace the lowest number with the next lowest number (the lowest number that remained in the batch after trimming). Thus 7.6 g becomes 9.1 g.

The new batch that results is a *Winsorized batch*. The Winsorized variance is calculated simply as the ordinary variance of this Winsorized batch.

Table 3.4. Calculating the 5% Trimmed Standard Deviation of Flake Weights from Pit 1 (Table 3.1)

Original batch x	Winsorized batch x_w	Deviation from mean $x_w - \bar{X}_w$	Squared deviation from mean $(x_w - \bar{X}_w)^2$
28.6	14.2	2.95	8.70
14.2	14.2	2.95	8.70
12.9	12.9	1.65	2.72
11.8	11.8	0.55	0.30
11.7	11.7	0.45	0.20
11.4	11.4	0.15	0.02
10.8	10.8	−0.45	0.20
10.5	10.5	−0.75	0.56
10.1	10.1	−1.15	1.32
9.2	9.2	−2.05	4.20
9.1	9.1	−2.15	4.62
7.6	9.1	−2.15	4.62

$$\bar{X}_w = 11.25 \text{ g} \qquad \Sigma(x_w - \bar{X}_w) = 0.00 \qquad \Sigma(x_w - \bar{X}_w)^2 = 36.16$$

$$\text{(sum of squares)}$$

$$s_w^2 = \frac{\Sigma(x_w - \bar{X}_w)^2}{n-1} = \frac{36.16}{11} = 3.29$$

$$s_T = \sqrt{\frac{(n-1)s_w^2}{n_T - 1}} = \sqrt{\frac{(12-1)3.29}{(10-1)}} = 2.01$$

These calculations are also shown in Table 3.4. (Notice that the mean involved in calculating the Winsorized variance is the mean of the Winsorized batch, which is not the same as the trimmed mean.) The trimmed standard deviation is derived from the Winsorized variance by the following equation:

$$s_T = \sqrt{\frac{(n-1)s_w^2}{n_T - 1}}$$

where

s_T is the trimmed standard deviation;
n is the number in the untrimmed batch;
s_w^2 is the variance of the Winsorized batch;
n_T is the number in the trimmed batch.

Table 3.4 also shows the calculation of the trimmed standard deviation for the flake weights from Pit 1. Comparison of the calculation columns for

---

Statpacks

Midspreads and standard deviations are pretty common fare in statpacks, and statpacks are truly helpful here because calculating a standard deviation with a calculator is time consuming (unless your calculator has a special key for doing it automatically). Trimmed standard deviations, however, are much less often provided for in statpacks. Just as in calculating a trimmed mean with your statpack, you are likely to have to adjust the batch yourself first. In this case, instead of replacing extreme values with missing data, you replace extreme values with the adjacent non-extreme value in the data. Once this modification has been made, the batch has been Winsorized, and the variance your statpack calculates on these numbers is the Winsorized variance, which you can convert into the trimmed standard deviation with your calculator.

Tables 3.3 and 3.4 shows quite clearly how the trimmed standard deviation avoids the overwhelming effect of outliers.

Just as the trimmed mean can be calculated for various trimming fractions, so can the trimmed standard deviation. In Chapter 2 we calculated a 25% trimmed mean of the flake weights from Pit 1 by trimming the three highest and the three lowest numbers from the batch. Calculation of the 25% trimmed standard deviation would begin with the creation of a Winsorized batch of 12 numbers in which the three highest numbers were replaced with the fourth highest and the three lowest numbers were replaced with the fourth lowest. From there on the calculation of the variance of the Winsorized batch and the trimmed standard deviation follow exactly the same path we have just taken for the 5% trimmed standard deviation.

WHICH INDEX TO USE

The range, the midspread, the standard deviation, and the trimmed standard deviation are all numerical indexes of the spread of a batch. Just as we asked when to use which index of the center of the batch, we must ask when to use which index of spread. The answer parallels that given in Chapter 2. The range is very widely understood but so badly affected by outliers that it is not often of much use. The midspread has been emphasized in exploratory data analysis. It is not as familiar as it should be to archaeologists, but it is easy to find and of wide utility for basic descriptive purposes. Its resistance to the effects of outliers makes it particularly attractive. The standard deviation is quite widely familiar (at least the term is, whether or not many archaeologists are really at home with the concept). Its statistical properties, like those of the mean, will serve us well in the rest of this book. It is of such importance that we will spend some effort on techniques to overcome its poor resistance to the effects of outliers. Some of these techniques are based on the trimmed standard deviation, and so we will be using the trimmed standard deviation when we work with batches whose outliers affect the standard deviation.

PRACTICE

Imagine you have conducted a regional survey of a small valley north of Nanxiong and have carefully measured the areas of the surface scatters

Table 3.5. Areas of Bronze Age Sites Near Nanxiong

Site area (ha)	
Early Bronze Age	Late Bronze Age
1.8	10.4
1.0	5.9
1.9	12.8
0.6	4.6
2.3	7.8
1.2	4.1
0.8	2.6
4.2	8.4
1.5	5.2
2.6	4.5
2.1	4.1
1.7	4.0
2.3	11.2
2.4	6.7
0.6	5.8
2.9	3.9
2.0	9.2
2.2	5.6
1.9	5.4
1.1	4.8
2.6	4.2
2.2	3.0
1.7	6.1
1.1	5.1
	6.3
	12.3
	3.9

that indicate the Bronze Age sites you encountered. The areas (in hectares) are given in Table 3.5.

1. Begin to explore these two batches of numbers with a back-to-back stem-and-leaf plot.
2. Continue your exploration by calculating the median, the mean, and the 10% trimmed mean for each batch and then the index of spread that corresponds to each of these indexes of level. Which pair of indexes makes most sense to use here? Why?
3. Based on the stem-and-leaf plots and the indexes of level and spread, what observations would you make about changes in site size from Early Bronze Age to Late Bronze Age near Nanxiong?

Comparing Batches

We have already compared batches with back-to-back stem-and-leaf plots, but there are quicker and more effective tools for graphically comparing batches. The numerical indexes of the center and spread of a batch that we have discussed in the last two chapters provide the basis for such tools. A standard way of plotting some of these indexes in exploratory data analysis is called the *box-and-dot* plot (or the *box-and-whisker* plot). The box-and-dot plot could, in theory, be based on any of the indexes of center and spread, but in practice the median and midspread are used.

THE BOX-AND-DOT PLOT

Construction of a box-and-dot plot begins in exactly the same way as construction of a stem-and-leaf plot: with the establishment of a scale along which the numbers in the batch will lie. Figure 4.1 presents a stem-and-leaf plot of post hole diameters from the Smith site (taken from Tables 1.7 and 1.8). To the right the stem is converted into a scale for drawing a box-and-dot plot. A horizontal line is placed next to 17.2 cm on this scale to represent the median. Two more lines at 18.8 cm and 15.7 cm represent the upper and lower quartiles, respectively. These three lines are framed with two vertical lines to form a box with a line across it near its center. This box graphically represents the midspread, that is, the central half of the numbers—those that

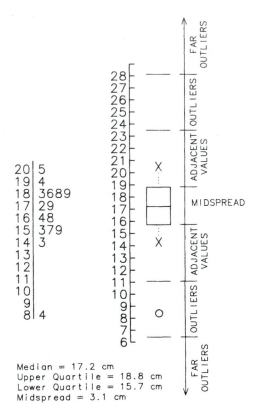

Figure 4.1. Box-and-dot plot of post hole diameters (in cm) from the Smith site.

fall between the two quartiles. The box provides a clear, clean picture of the most important central bunch of numbers in the batch, one that is more quickly perceived than the stem-and-leaf plot.

We can include more detail in the box-and-dot plot, and at the same time provide more precise definition of other important features of a batch. We have, for example, already discussed outliers, numbers that fall far outside the central bunch and are generally a nuisance as far as several otherwise very useful numerical indexes of center and spread are concerned. It is often quite helpful to simply eliminate outliers, but we are frequently confronted with borderline cases—numbers that lie beyond the central bunch but not so far outside it as to make us certain that they do not belong with the batch.

The box-and-dot plot provides a graphical approach to identifying outliers consistently and signaling their presence by suggesting a rule of thumb to distinguish between the central bunch of numbers and outliers. According

Rules of Thumb

Practical statistics is filled with rules of thumb that are efforts to patch the gaps between nice neat principles and much messier real life. Outliers create just such a gap. We considered the case of Edelbert Pea (the boss's nephew) as an example of an outlier. He was easy to identify as an outlier because he made $52.00 an hour while all the other shipping clerks made $8.00 an hour. But what if Edelbert made only $8.50 an hour? And what if he had worked at L.L. Pea for three years as a shipping clerk, while all the other shipping clerks who made $8.00 an hour had less than six months' experience? He no longer seems such an outlier. In fact, he begins to sound like a good example of exactly the kind of variation we would like to include in our study of shipping clerks' wages. But where would we draw the line? If Edelbert made $12.00 an hour would he be an outlier? If he made $20.00? In seeking to draw a line at the point where Edelbert's wages make him an outlier, we're basically trying to do the impossible. At some level it is just silly to pretend to be able to say, for example, that if he made as much as $14.73 an hour he would not be an outlier, but if he made $14.74 he would be. The judgment is just much fuzzier than that. On the other hand, if we're going to analyze shipping clerks' wages we either have to include Edelbert or exclude him. There is no middle ground. "Maybe" simply does not lead to any course of action we can pursue. It is in precisely such situations that statisticians make up rules of thumb—to provide systematic guidance where the best answer is "maybe" but the only useful answers are "yes" and "no."

Saying that a number is an outlier if it falls more than 1.5 midspreads outside either quartile of its batch is a rule of thumb. It provides us with a systematic way of identifying outliers in a batch according to a clearcut rule. But it would be hard to justify choosing exactly 1.5 midspreads rather than 1.6 or 1.4 because the choice, finally, is somewhat arbitrary. Indeed, there is some variation from one statistics book (or one computer program) to the next in the exact rule of thumb used for identifying outliers. The same is true of the other rules of thumb that we will discuss as we go on through this book.

---

to this rule of thumb, an outlier is any number that lies more than one and a half times the length of the box beyond either end of the box. We can think of this in purely graphical terms. We could measure the box in a box-and-dot plot as drawn on paper. If the box is 1 inch long, then we would say that any number falling more than 1.5 inches above the top of the box or below the bottom of the box is an outlier. Further, any number falling more than twice this far from the end of the box is a *far outlier*. These distances are indicated with lines in Figure 4.1.

The same result can be achieved mathematically. Since the length of the box is the midspread, the distance that defines outliers is 1.5 times the midspread ($1.5 \times 3.1 = 4.65$ cm in the example in Figure 4.1). Since the top of the box represents the upper quartile, the position of the defining line for high outliers on the number scale is the upper quartile plus 1.5 times the midspread ($18.8 + 4.65 = 23.45$ cm in the example in Figure 4.1). Since the bottom of the box represents the lower quartile, the position of the defining line for low outliers is the lower quartile minus 1.5 times the midspread ($15.7 - 4.65 = 11.05$ cm in the example in Figure 4.1).

In the same way, the positions of the defining lines for far outliers can be established mathematically. The line defining far high outliers is twice as far above the upper quartile as the line defining outliers. That is to say, instead of 1.5 times the midspread beyond the quartiles, the far outlier defining line falls at 3 times the midspread beyond the quartiles ($18.8 + 9.3 = 28.1$ cm and $15.7 - 9.3 = 6.4$ cm in Figure 4.1).

Thus the areas above and below the box in the box-and-dot plot are each divided into three zones. Numbers that fall in the nearest zone above or below the box are called *adjacent values*. These numbers are outside the central half of the batch but are still considered part of the main bunch of numbers. In the next zone away from the median come outliers, and in the farthest zone are far outliers. Ordinarily these zones are not indicated by lines the way they are in Figure 4.1. Instead, they are distinguished by different symbols representing the numbers that fall in them. The highest and lowest adjacent values are indicated with X's, as shown in Figure 4.1. These X's, then, represent the extremes of the main bunch of numbers (excluding all outliers). Outliers are all indicated individually on the plot as hollow dots, and far outliers are all indicated individually as solid dots. The batch represented in Figure 4.1 has only one outlier (8.4 cm) and no far outliers, so there is a single hollow dot and no solid dots. These conventions about X's, hollow dots, and solid dots stand for the labels and lines drawn to the right of Figure 4.1, so such labels and defining lines do not generally appear when box-and-dot plots are drawn. As is the case with rules of thumb, the exact conventions used to indicate outliers and far outliers in box-and-dot plots vary from one book or program to the next.

The box-and-dot plot makes it easy to compare several batches. In Chapter 1, we compared the batch used for the example in Figure 4.1 to another batch of post hole diameters with a back-to-back stem-and-leaf plot (Table 1.8). Figure 4.2 compares the same two batches with two box-and-dot plots instead. The box-and-dot plot for post hole diameters at the Smith site is exactly the same as in Figure 4.1 (except that it is now on a longer scale). The box-and-dot plot for post hole diameters at the Black site is made in exactly the same manner, but using the numbers listed in Table 1.7 for the Black site. The one extremely large post hole qualifies as a far outlier, since it

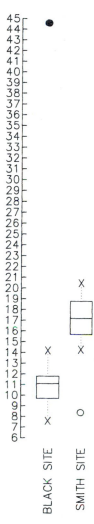

Figure 4.2. Box-and-dot plots comparing post hole diameters (in cm) at the Black–Smith sites.

lies more than 3 times the length of the box from the box's upper end. It is thus shown as a solid dot.

When we look at the box-and-dot plots in Figure 4.2, we quickly reach the same conclusion we reached looking at the back-to-back stem-and-leaf plot of these same numbers in Table 1.8. At each site there is a post hole that does not seem to represent the same kind of phenomenon as the rest of the post holes—an extremely large post hole at the Black site and an extremely small post hole at the Smith site. In general, post holes at the Smith site are larger than post holes at the Black site by a margin of 5 or 6 cm. The box-and-dot plot shows us these patterns even more clearly than the back-to-back stem-and-leaf plot because the box-and-dot plot is a simpler, more quickly perceived way of representing the basic features of each batch. The box-and-dot plot can also be extended easily to the comparison of a larger number of separate batches simply by adding additional boxes and dots to the same scale. The back-to-back stem-and-leaf plot cannot be extended very conveniently to the comparison of more than two batches.

REMOVING THE LEVEL

When we compare two or more batches of numbers, as in Figure 4.2, probably the most noticeable characteristic of each batch is its level, or center. If we want to compare other features of the batches, it is convenient to remove the conspicuous effect of their differing levels. We do this by reducing the levels of both batches to zero.

Figure 4.3 shows this graphically. We have simply slid both box-and-dot plots down the scale so that the center of each (that is, the median) lines up with the zero point on the scale. The same result can also be achieved mathematically by subtracting the median of a batch from each number in the batch. For example, we take all the post hole diameters from the Smith site in Table 1.7, and subtract the median of Smith site post hole diameters (17.2 cm) from each one, as shown in Table 4.1. The result is a new set of numbers that represent how much each post hole is larger or smaller than the median size. Post holes whose diameters are larger than the median are represented by positive numbers and post holes whose diameters are smaller than the median are represented by negative numbers. We could arrive at the Smith site box-and-dot plot with the level removed (in Figure 4.3) by making a box-and-dot plot of this new batch of numbers. The result would be exactly the same as graphically sliding the box-and-dot plot made previously down the scale until its median arrived at zero. (If you do not immediately see why this is so, the best way to understand is to try it out for yourself.)

Having removed the levels from these two batches of numbers we can no longer compare them in regard to level. The process of removing the level is to artificially set the center of both batches at zero. With the conspicuous effect of differences in level removed, however, we very quickly notice that the two batches differ in regard to spread. Disregarding the outliers and far outliers, we see that the adjacent values in both batches are similarly spread out on the number scale. The most central bunch of numbers, however (the middle half as represented by the box), is more spread out for the Smith site post holes than for the Black site post holes. This difference was certainly

Figure 4.3. Box-and-dot plots of post hole diameters (in cm) at the Black–Smith sites with levels removed.

Table 4.1. Removing the Level from Smith Site Post Hole Diameters by Subtracting the Median (17.2 cm)

20.5^a	–	17.2	=	3.3		19.4	–	17.2	=	2.2
17.2	–	17.2	=	0.0		16.4	–	17.2	=	–0.8
15.3	–	17.2	=	–1.9		18.8	–	17.2	=	1.6
15.9	–	17.2	=	–1.3		15.7	–	17.2	=	–1.5
18.3	–	17.2	=	1.1		18.9	–	17.2	=	1.7
17.9	–	17.2	=	0.7		16.8	–	17.2	=	–0.4
18.6	–	17.2	=	1.4		8.4	–	17.2	=	–8.8
14.3	–	17.2	=	–2.9						

^{a}All values in centimeters.

visible in the previous box-and-dot plot (Figure 4.2), but it is considerably more conspicuous now that the two boxes have been lined up at their middles by removing the levels.

REMOVING THE SPREAD

Just as we removed the level from a batch by reducing its center to zero, we can remove the spread from a batch by reducing its spread to one. This must be accomplished mathematically; it cannot be done graphically, as in the case of removing the level by sliding the box down the number scale. Once the level has been removed mathematically, however, by subtracting the median, we remove the spread by dividing by the midspread. Table 4.2 continues where the calculations in Table 4.1 left off with the Smith site post hole diameters. For example, the first number in the batch in Table 4.1 represents a post hole 20.5 cm in diameter. When the level is removed from

Table 4.2. Removing the Spread from Smith Site Post Hole Diameters by Dividing by the Midspread (3.1 cm) after the Level Has Been Removed (Compare to Table 4.1)

3.3^a	/	3.1	=	1.06		2.2	/	3.1	=	0.71
0.0	/	3.1	=	0.00		–0.8	/	3.1	=	–0.26
–1.9	/	3.1	=	–0.61		1.6	/	3.1	=	0.52
–1.3	/	3.1	=	–0.42		–1.5	/	3.1	=	–0.48
1.1	/	3.1	=	0.35		1.7	/	3.1	=	0.55
0.7	/	3.1	=	0.23		–0.4	/	3.1	=	–0.13
1.4	/	3.1	=	0.45		–8.8	/	3.1	=	–2.84
–2.9	/	3.1	=	–0.94						

^{a}All values in centimeters.

Statpacks

As in the case of stem-and-leaf plots, there are many statistical computer programs that draw box-and-dot plots. Their conventions for indicating outliers may vary from those used in this book, but as long as you know what they are, that should not pose a problem. Some programs draw box-and-dot plots vertically, the way they are drawn in this book, although the lower numbers tend to be higher on the screen, and the higher numbers tend to be lower on the screen in contrast to the figures here. Some programs draw the plots horizontally. None of this makes any difference, of course, to the interpretation of the plots. Usually such programs automatically choose a scale for the plots, releasing your time and energy for other more important tasks. If your program does not automatically produce box-and-dot plots of several batches all at the same scale for comparing several batches, however, you may need to look up in the manual how to take active control of determining the scale to be used. Clearly, box-and-dot plots of different batches cannot be compared to each other unless they are drawn to the same scale.

The easiest way to make box-and-dot plots with the level or level and spread removed, of course, is with a statistics computer package. Usually the procedure you need to follow is to *transform* the numbers in the original batch by subtracting the median from each (and dividing the result by the midspread if you want to remove the level and the spread) to create a new batch (or *variable*). Almost all statpacks make it easy to do such a thing. Then you can make a box-and-dot plot of the new batch.

---

this number, we see that this post hole diameter is 3.3 cm larger than the median. Continuing with the calculation in Table 4.2, we see that this 3.3 cm divided by 3.1 cm (the midspread) is 1.06. This result, 1.06, means that the post hole diameter in question is above the median by an amount equal to a little bit more than one midspread. In the box-and-dot plot (Figure 4.1), this post hole would lie about the length of the box above the median (that is, above the center line of the box). Since this post hole provides the highest adjacent value, it is, in fact, located in Figure 4.1 as the X above the box, and the center of this X does, indeed, lie about the length of the box above the box's center line.

Removing the level and spread from both batches of numbers—the post hole diameters from the Black site and from the Smith site—and making yet another box-and-dot plot of the result gives us Figure 4.4. The centers of both batches are still at zero, but now the boxes representing the middle half of each batch are the same length. That length, of course, is one, since the box length always represents the midspread, and removing the level and

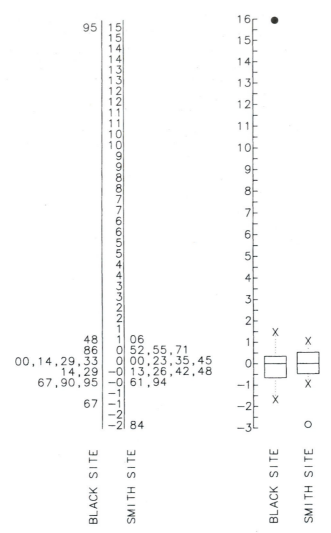

Figure 4.4. Stem-and-leaf and box-and-dot plots of post hole diameters at the Black–Smith sites with levels and spreads removed.

spread has the effect of setting the center at zero and the spread at one. We thus cannot use Figure 4.4 to compare the batches in regard to either level or spread. The feature that becomes most conspicuous at this point is shape, especially symmetry. Clearly, the post hole diameters from the Black site tend to spread out downward from the median more than upward. Remember that one-quarter of the numbers in the batch fall between the median and the top of the box and one-quarter fall between the median and the bottom of the box. The one-quarter of the numbers immediately below the median at the Black site are clearly more spread out than the one-quarter immediately above the median, which clump closer to the median. The Smith site post hole diameters, on the other hand, have a more symmetrical distribution, although its middle half spreads upward a little more than it does downward. We will discuss shape and symmetry in more detail in the next chapter.

It is worth noting that there is an easier way to draw a box-and-dot plot with the level and spread removed. We have just subtracted the median from all the numbers in the batch and divided all the resulting numbers by the midspread to arrive at a new batch. In this batch we found the median, upper and lower quartiles, outliers, etc., so as to draw a new box-and-dot plot from scratch. We simply could have applied this treatment to each of the five numbers required to define the box-and-dot plot (the median, the upper and lower quartiles, and the upper and lower extreme adjacent values). These five values, with the level and spread removed, produce the same box-and-dot plot as the same five values determined afresh from a complete new batch with the level and spread removed from all the numbers. To finish the graph requires only subtracting the median from each outlier and dividing the result by the midspread so as to locate outliers on the new number scale.

UNUSUALNESS

This new number scale is a very interesting one. It is no longer a scale of centimeters as the previous number scales have been, but rather, in effect, a scale of unusualness. It locates each number in the batch according to just how central or how peripheral that number is in terms of the batch to which it belongs. Unusualness is not an inherent property of a thing but rather a statement of how a thing relates to the group of which it is a member. If a thing falls well within the central bunch of things in its group, then it is not very unusual. If a thing falls in a more peripheral position, relative to the central bunch of things in its group, then it is more unusual. In a group of professional basketball all-stars, a person 6'-6" tall is not very unusual. In a group of university professors, however, a person 6'-6" tall is very unusual. Removing the level and spread from a batch of numbers gives us a scale along which we can express unusualness in a standard and systematic manner. For this reason the traditional statistical term for this procedure is *standardizing*.

The number scale in Figure 4.4 expresses how far each number in each batch departs from the median for that batch in terms of the midspread for that batch. The first number in Table 4.1, for example, is 20.5 cm, which represents a post hole 3.3 cm larger than the median diameter for the Smith

Table 4.3. Removing the Level and Spread from Black Site Post Hole Diameters by Subtracting the Median (11.1 cm) and Dividing by the Midspread (2.1 cm)

(9.7[a] – 11.1)	/	2.1	= –0.67	(11.7 – 11.1)	/ 2.1	=	0.29
(9.2 – 11.1)	/	2.1	= –0.90	(11.1 – 11.1)	/ 2.1	=	0.00
(12.9 – 11.1)	/	2.1	= 0.86	(7.6 – 11.1)	/ 2.1	=	–1.67
(11.4 – 11.1)	/	2.1	= 0.14	(11.8 – 11.1)	/ 2.1	=	0.33
(9.1 – 11.1)	/	2.1	= –0.95	(14.2 – 11.1)	/ 2.1	=	1.48
(44.6 – 11.1)	/	2.1	= 15.95	(10.8 – 11.1)	/ 2.1	=	–0.14
(10.5 – 11.1)	/	2.1	= –0.29				

[a]All values in centimeters.

site. This post hole measurement becomes 1.06 in the standardized batch (Table 4.2), meaning that its diameter is slightly more than 1 midspread above the median. The second measurement in Table 4.1, 17.2 cm, is the median. Thus its difference from the median is 0.0 cm or 0.00 midspreads. The third measurement, 15.3 cm, falls 1.9 cm below the median, and becomes –0.61 in the standardized batch. It is thus 0.61 midspreads below the median value. The first post hole (20.5 cm in diameter), then, is more unusual than the third post hole, because it falls farther out toward the periphery of the batch.

This standardized number scale permits comparisons of unusualness from one batch to another. For example, the first post hole in Table 4.3, with a diameter of 9.7 cm, is 1.4 cm smaller than the median diameter at the Black site. The 15.7-cm post hole at the Smith site (fourth from the bottom in Tables 4.1 and 4.2) is 1.5 cm smaller than the median diameter in its batch. It might seem that this latter post hole is more unusual since it is farther from the center of its batch in centimeters. It is, however, in a batch that is just more spread out in general. At the Black site a post hole 1.4 cm smaller than the median lies 0.67 midspreads away from the median. At the Smith site a post hole 1.5 cm smaller than the median lies only 0.48 midspreads away from the median. The post hole with a diameter of 15.7 cm is thus more unusual for the Smith site than is the post hole with a diameter of 9.7 cm for the Black site.

Perhaps the context in which we most frequently encounter such unusualness scales is in standardized testing. Elementary school test results are often expressed in terms of how far above or below average a score is for a particular grade level. The *percentiles* in which college entrance examinations are commonly expressed also provide such information. A student who scores in the 75th percentile knows that about 75% of those taking the test had lower scores, while about 25% had higher scores. If the batch in question is symmetrical, the 75th percentile is equivalent to a score of about .5 on the unusualness scale we have been discussing. This is because when a batch is standardized by subtracting the median and dividing by the midspread, a score of .5 means above the median by half a midspread. A number that is half a midspread above the median is, of course, the upper quartile (at least in a symmetrical batch). And the upper quartile is the number above which lie 25% of the numbers in the batch.

STANDARDIZING BASED ON THE MEAN AND STANDARD DEVIATION

Expressing the unusualness of a number in terms of its centrality or peripheralness in its own batch is a critically important concept in statistics. Much of the remainder of this book is built on this concept of unusualness. In this chapter we have focused on removing the level and spread using the median and midspread as the numerical indexes of level and spread. We have done this because the box-and-dot plot based on the median and midspread provides a particularly easy graphical illustration of the procedure and its implications. It is more common and ultimately much more useful to use the mean and standard deviation, however, because these indexes have some especially attractive mathematical properties. The basic principles and the calculations are an exact parallel to what we have just discussed. To standardize a batch using the mean and standard deviation, you subtract the mean of the batch from every number in the batch and divide the result by the standard deviation of the batch. The resulting batch is often referred to as a batch of *standard scores*, or *z scores*. The *z* scores tell how many standard deviations above the mean (for positive *z* scores) or below the mean (for negative *z* scores) each number in the original batch falls.

PRACTICE

1. Continue to explore the areas of sites near Nanxiong given in Table 3.5 by making box-and-dot plots of Early and Late Bronze Age site areas. How do the levels of the two batches compare?
2. Now compare Early and Late Bronze Age site areas by drawing box-and-dot plots with the levels removed. How do the spreads of the two batches compare?
3. Now compare Early and Late Bronze Age site areas by drawing box-and-dot plots with the levels and spreads removed. How do the two batches compare in terms of symmetry?

4. The largest Early Bronze Age site is 4.2 ha; the largest Late Bronze Age site is 12.8 ha. Which of these sites is more unusual in terms of its batch? Why? Use the median and the midspread of each batch to provide a score for the unusualness of each of these sites. Use the mean and standard deviation of each batch to do the same thing. Do these scores confirm your assessment of which site is more unusual in its batch?

Chapter 5

The Shape, or Distribution, of a Batch

The *shape* of a batch refers to the way in which the numbers are distributed along the number scale, apart from level and spread. The traditional statistical term for shape is *distribution*. There are two principal aspects to the shape of a batch: *number of peaks* and *symmetry*. We have already discussed batches with multiple peaks and some of the reasons why they must be divided before analysis, so we will proceed directly to the second aspect.

SYMMETRY

Once we have a batch with a single peak, we are in position to use numerical indexes of level and spread. One use to which we can put these numerical indexes is in removing the level and spread so as to evaluate symmetry more carefully. A batch may be symmetrically distributed about its single peak. In a symmetrical batch, about half the numbers fall above the peak, about half the numbers fall below the peak, *and* the numbers above and below the peak stretch away from the peak to similar degrees. That is, the numbers on one side of the peak are no more closely bunched up near the peak than are those on the other side of the peak.

Table 5.1 lists a batch of measured volumes of bell-shaped storage pits and illustrates the symmetry of the batch with a stem-and-leaf plot. The

**Table 5.1. Volumes of Bell-Shaped
Storage Pits at the Buena Vista Site**

Volumes (m^3)	Stem-and-leaf plot	
1.23	16	5
1.48	15	15
1.55	14	0568
1.38	13	24589
1.10	12	1349
1.02	11	02
1.29	10	2
1.32		
1.35		
1.65		
1.39		
1.40		
1.12		
1.46		
1.24		
1.34		
1.21		
1.45		
1.51		

stem-and-leaf plot, in fact, shows perfect symmetry. The distribution of numbers above the peak is a mirror image of the distribution of numbers below the peak. The median of this batch is 1.35 m^3, and its mean is 1.34 m^3. Such close agreement between median and mean is characteristic of batches with symmetrical distributions, and both these numerical indexes of level fall right at the central peak on the stem-and-leaf plot. In short, both behave very well in a symmetrical single-peaked batch. They give us exactly the index of the center that matches the pattern that is so clear in the stem-and-leaf plot.

It is unusual to find the perfect symmetry of Table 5.1 in real-world batches of numbers, especially in such small batches as this one. We would be willing to accept a batch this small as symmetrical even if the pattern were considerably less than perfect. Judgments about symmetry are subjective, and we will discuss the process of making them more fully below.

Table 5.2 lists another batch of measured volumes of bell-shaped storage pits from a different site. As the stem-and-leaf plot shows, however, this batch is not nearly so symmetrical. Most of the numbers are above the peak, and they tend to stray far above the peak. In contrast, the numbers below the peak are few and lie quite close to the peak. This is an *asymmetrical*, or *skewed*, distribution. Batches can be skewed upward as this one is, or downward if the values tend to stray toward the lower numbers. For discussing

**Table 5.2. Volumes of Bell-Shaped
Storage Pits at the Buenos Aires Site**

Volumes (m^3)	Stem-and-leaf plot	
1.22	20	3
1.64	19	
1.16	18	4
1.07	17	
1.50	16	4
1.84	15	0
1.37	14	03
1.15	13	27
1.29	12	269
1.32	11	1567
2.03	10	47
1.17		
1.04		
1.43		
1.11		
1.40		
1.26		

symmetry it is especially convenient to draw stem-and-leaf plots with lower numbers at the bottom and higher numbers at the top—like the ones in this book—so that the values in an upwardly skewed shape stray upward on the plot. If your statpack draws them the other way, just remember that when we talk about upward skewness we mean a shape that strays toward the higher numbers, not necessarily toward the top of the stem-and-leaf plot.

Numerical indexes of the center do not behave well at all for a skewed distribution. The median for this batch is 1.29 m^3, and the mean is 1.35 m^3. These two indexes differ more than did the median and mean for the batch in Table 5.1. More important, both fall too high on the number scale to accurately reflect the clear single peak at about 1.1 m^3. The effect of a skewed distribution on numerical indexes is quite similar to the effect of outliers, as discussed in Chapter 2. Indeed, it is sometimes difficult to tell whether we are looking at a stem-and-leaf plot containing outliers or one showing a skewed distribution. Even the median, highly resistant to the effects of outliers, is affected by a skewed distribution since skewing consists not just of a few aberrant measurements but rather of a pervasive tendency in the shape of the batch.

Since we need a numerical index of the level and spread of a batch in order to begin virtually any statistical analysis, such asymmetrical shapes present us with a serious impediment. Sometimes using the trimmed mean and trimmed standard deviation can help, but it is really the effect of outliers that these indexes eliminate nicely. More fundamental remedies are usually called for before working with a badly asymmetrical shape.

TRANSFORMATIONS

We have already seen that we can perform at least some kinds of arithmetic operations on all the numbers in a batch to produce a new batch that is more amenable to certain kinds of examination. For example, we subtracted the median or the mean from all the numbers in a batch to produce a new batch with the level removed. This had the effect of setting the center to a standard value (zero), while the spread and shape of the batch remained the same. Then we divided all numbers in the zero-level batch by the midspread or the standard deviation to remove the spread. This had the effect of setting the spread to a standard value of one, while the shape of the batch remained the same. *Transformations* are a way of removing the shape of a batch or setting it to a standard shape (single-peaked and symmetrical).

The operations of removing the level and spread are related to each other: first we remove the level, then the spread. We do not remove the spread from a batch without removing the level first. Transformations of shape, however, are independent of removing level and spread. Such transformations are often performed on batches without removing the level or spread, although they can also be applied after removal of level and spread. Figure 5.1 illustrates the effects that several commonly used transformations have on the shape of a batch. Each batch of numbers is accompanied by a stem-and-leaf plot and by a box-and-dot plot with the level and spread removed. The box-and-dot plots provide the most sensitive indication of symmetry in the original batch and in its various transformations.

Looking first at the original batch of numbers (x in Figure 5.1), the stem-and-leaf plot shows perfect symmetry (as it did in Table 5.1). The box-and-dot plot confirms this impression.

The transformed batch in the second column of Figure 5.1 is produced by taking the square root of each of the numbers in the original batch (see, for example, the first number: $\sqrt{1.230} = 1.109$). This is commonly referred to as the *square root transformation*. The stem-and-leaf plot and the box-and-dot plot for the square root transformation reveal that this new batch has a recognizable tendency to stray downward from its center. (Compare the midspread box or the two extreme adjacent values to those of the original batch in the box-and-dot plots.) The effect of the square root transformation is always to produce a new batch more strongly skewed downward than the original batch, just as we see in this case.

Logarithms

The logarithm of a number is the power to which some base must be raised to produce the number. For example, the base-10 logarithm of 1000 is 3 since 10^3 = 1000. The base-10 logarithm of 100 is 2, since 10^2 = 100. The base-10 logarithm of 10 is 1, since 10^1 = 10. We do not usually raise numbers to fractional powers in simple mathematics, but it can be done. Since $10^2 = 100$ and 10^3 = 1000, $10^{2.14}$ must be greater than 100 and less than 1000. In fact, $10^{2.14}$ = 137.2, so the base-10 logarithm of 137.2 is 2.14. One of the vexing chores of introductory statistics used to be the technique of using a table of logarithms. Fortunately, computers have made logarithm tables pretty much obsolete for transforming batches, and we can now assume that logarithm transformations will be done with computers or at least calculators.

The numbers of the third column of Figure 5.1 are actually *natural logarithms*, or base *e* logarithms. The mathematical constant *e* has a value of approximately 2.7182818. Its useful characteristics in theoretical mathematics are not of importance to us here, but the logarithms used in many statpacks are base-*e* logarithms. Thus the numbers in the third column are the powers to which *e* must be raised to produce the numbers in the original batch. The first number in the original batch, for example, is 1.230. Since $2.7182818^{.207}$ = 1.230, the natural logarithm of 1.230 is .207, and .207 appears first in the third column.

The transformed batch in the third column of Figure 5.1 is produced by finding the logarithm of each of the numbers in the original batch. As the box-and-dot plots show, this *logarithm transformation*, or log transformation for short, is skewed downward even more strongly than is the square root transformation. Like the square root transformation, it produces a batch with a more downward skewness than the original batch. The effect of the log transformation in this regard is even stronger than the effect of the square root transformation.

The transformed batch listed in the fourth column of Figure 5.1 is produced with the *negative reciprocal transformation* ($-1/x$). The negative reciprocal of the first number in the batch (1.230) is $-1/1.230 = -0.813$. Like the other transformations discussed above, it produces a transformed batch with a more pronounced downward skewing than the original batch. Its effect is even stronger than that of the other transformations, as can be seen in the box-and-dot plots at the bottom of Figure 5.1.

The fifth column of Figure 5.1 shows an even stronger effect in the same direction. This transformation ($-1/x^2$) produces downward skewness to an even greater degree. Using the first number again, as an example of the calculation, $(-1/1.230^2) = -0.661$. We could continue this progression indefinitely with transformations creating stronger and stronger downward skewness: $-1/x^3$, $-1/x^4$, etc.

Beginning in the sixth column, Figure 5.1 illustrates transformations that produce the opposite effect. The *square transformation* is simply x^2. (For

x	$\sqrt{x}$	$\log(x)$	$-\dfrac{1}{x}$	$-\dfrac{1}{x^2}$	x^2	x^3	x^4
1.230	1.109	0.207	−0.813	−0.661	1.513	1.861	2.289
1.480	1.217	0.392	−0.676	−0.457	2.190	3.242	4.798
1.550	1.245	0.438	−0.645	−0.416	2.403	3.724	5.772
1.380	1.175	0.322	−0.725	−0.525	1.904	2.628	3.627
1.100	1.049	0.095	−0.909	−0.826	1.210	1.331	1.464
1.020	1.010	0.020	−0.980	−0.961	1.040	1.061	1.082
1.290	1.136	0.255	−0.775	−0.601	1.664	2.147	2.769
1.320	1.149	0.278	−0.758	−0.574	1.742	2.300	3.036
1.350	1.162	0.300	−0.741	−0.549	1.823	2.460	3.322
1.650	1.285	0.501	−0.606	−0.367	2.723	4.492	7.412
1.390	1.179	0.329	−0.719	−0.518	1.932	2.686	3.733
1.400	1.183	0.336	−0.714	−0.510	1.960	2.744	3.842
1.120	1.058	0.113	−0.893	−0.797	1.254	1.405	1.574
1.460	1.208	0.378	−0.685	−0.469	2.132	3.112	4.544
1.240	1.114	0.215	−0.806	−0.650	1.538	1.907	2.364
1.340	1.158	0.293	−0.746	−0.557	1.796	2.406	3.224
1.210	1.100	0.191	−0.826	−0.683	1.464	1.772	2.144
1.450	1.204	0.372	−0.690	−0.476	2.103	3.049	4.421
1.510	1.229	0.412	−0.662	−0.439	2.280	3.443	5.199

Stem-and-leaf plots:

```
16 | 5        12 | 59      5 | 0        -6 | 1        -3 | 7       2 | 7        4 | 5        7 | 4
15 | 15       12 | 0123    4 | 14       -6 | 99865   -4 | 87642   2 | 4        3 | 7        6 |
14 | 0568     11 | 566888  3 | 0234789  -7 | 4321    -5 | 765321  2 | 23       3 | 0124     5 | 28
13 | 24589    11 | 0114    2 | 12689    -7 | 865     -6 | 8650    2 | 011      2 | 5677     4 | 458
12 | 1349     10 | 56      1 | 019      -8 | 311     -7 |         1 | 8899     2 | 134      3 | 023678
11 | 02       10 | 1       0 | 2        -8 | 9       -8 | 30      1 | 77       1 | 899      2 | 1348
10 | 2                                  -9 | 1       -9 | 6       1 | 555      1 | 134      1 | 156
                                        -9 | 8                    1 | 23
                                                                  1 | 0
```

Box-and-dot plots with levels and spreads removed:

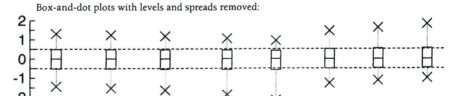

Figure 5.1. The effect of transformations on the shape of the batch of measurements from Table 5.1.

the first number in the batch, $1.230^2 = 1.513$.) The upward straying effect of the numbers in this small batch after applying the square transformation is barely noticeable.

The *cube transformation* in the seventh column, however, is stronger, and the upward straying of numbers in this transformed batch is easily recognized in the box-and-dot plot. The calculation in this case is simply to raise the original number to the next higher power than in the previous transformation. (For the first number, $1.230^3 = 1.861$.) Even stronger, and in the same positive direction, is the skewing effect of the x^4 transformation in the last column of Figure 5.1. As with the earlier sequence of transformations producing downward skewing, we could continue this sequence indefinitely to higher and higher powers.

CORRECTING ASYMMETRY

We have just seen how a series of transformations can change the shape of a batch. In this example we started with a batch the shape of which was already symmetrical, and progressively skewed it farther and farther, first in the downward direction, and then in the upward direction. Once we understand them, the effects of the various transformations can be put to good use in changing the shapes of batches that are difficult to work with because their distributions are not symmetrical. When a transformation producing upward skewing is applied to a batch with downward skewness, the result may be a symmetrical shape.

Precisely the transformations we have just discussed are often used to "correct" asymmetrical shapes. We can use the experience gained in the previous example to list these common transformations and their effects. Table 5.3 summarizes the experience gained from examining the graphs in Figure 5.1. Or, to put it another way, Figure 5.1 graphs the practical impact

Table 5.3. Transformations for Correcting Asymmetry

x^4	Stronger effect	
x^3	Strong effect	Produce upward skewness, that is, correct downward skewness
x^2	Mild effect	
x	No effect	
$\sqrt{x}$	Weak effect	
$\log(x)$	Mild effect	Produce downward skewness, that is, correct upward skewness
$\dfrac{1}{x}$	Strong effect	
$-\dfrac{1}{x^2}$	Stronger effect	

of applying the transformations listed in Table 5.3. Table 5.3 can be used to select an appropriate transformation to apply to an asymmetrical batch like the one in Table 5.2. This batch has a very pronounced tendency to stray upward, so we will need one of the transformations from the lower half of Table 5.3—transformations that correct upward skewness. The effects of all four of these transformations are illustrated in Figure 5.2.

We easily identified the shape of the batch in the stem-and-leaf plot in Table 5.2 as upwardly skewed. It may be difficult, however, to decide whether a few numbers that stray far from the central bunch represent an upwardly skewed distribution or genuine outliers. The rules of thumb for identifying outliers in box-and-dot plots label the highest two values in the original batch as outliers, as can be seen in the first column of Figure 5.2. Nevertheless, these rules of thumb, as discussed in Chapter 4, are only arbitrary ways to simplify a complicated relationship between straying numbers and the batch of which they may or may not be a meaningful part. Another approach is to see what effect transformations have on possible outliers.

The weakest transformation for correcting upward straying is the square root transformation, illustrated in the second column of Figure 5.2. In the transformed batch, the nearer of the two outliers in the untransformed batch no longer qualifies as an outlier, and the box representing the midspread comes closer to being centered on the median. Even disregarding the one outlier still identified, the adjacent values clearly stray farther up than down. The square root transformation produced a less asymmetrical batch, but stronger action is necessary.

The next stronger transformation is the log transformation, illustrated in the third column of Figure 5.2. In this batch, the median is very close to the center of the midspread. The highest value is still identified as an outlier, but disregarding it, the adjacent values still stray considerably farther upward than downward. The stem-and-leaf plot shows this upward skewness quite clearly. A still stronger transformation is at least worth trying in this instance.

The fourth column in Figure 5.2 illustrates the effect of the negative reciprocal transformation. The midspread box has now slipped below the middle of the number scale, but the adjacent values still stray farther upward than downward. Most conspicuous, the last remaining outlier no longer qualifies for that status according to the usual rules of thumb. When outliers disappear under the effect of transformations that are also improving the general symmetry of the distribution, it is an indication that they should not be eliminated as outliers but rather treated as straying members of the batch. In such cases, the use of an appropriate transformation is a preferable treatment to correct both asymmetry and apparent outliers. Since the adjacent values continue to stray upward in such a pronounced fashion, it is worth investigating one more transformation with a yet stronger effect.

x	$\sqrt{x}$	$\log(x)$	$-\dfrac{1}{x}$	$-\dfrac{1}{x^2}$
1.220	1.105	0.199	−0.820	−0.672
1.640	1.281	0.495	−0.610	−0.372
1.160	1.077	0.148	−0.862	−0.743
1.070	1.034	0.068	−0.935	−0.873
1.500	1.225	0.405	−0.667	−0.444
1.840	1.356	0.610	−0.543	−0.295
1.370	1.170	0.315	−0.730	−0.533
1.150	1.072	0.140	−0.870	−0.756
1.290	1.136	0.255	−0.775	−0.601
1.320	1.149	0.278	−0.758	−0.574
2.030	1.425	0.708	−0.493	−0.243
1.170	1.082	0.157	−0.855	−0.731
1.040	1.020	0.039	−0.962	−0.925
1.430	1.196	0.358	−0.699	−0.489
1.110	1.054	0.104	−0.901	−0.812
1.400	1.183	0.336	−0.714	−0.510
1.260	1.122	0.231	−0.794	−0.630

	x	$\sqrt{x}$	$\log(x)$	$-\dfrac{1}{x}$	$-\dfrac{1}{x^2}$
Mean:	1.353	1.158	0.285	−0.764	−0.600
Median:	1.290	1.136	0.255	−0.775	−0.601

Stem-and-leaf plots:

```
20 | 3        14 | 3        7 | 1       −4 | 9        −2 | 4
19 |          13 | 6        6 | 1       −5 | 4        −3 | 70
18 | 4        13 |          5 | 0       −6 | 71       −4 | 94
17 |          12 | 8        4 | 1       −7 | 986310   −5 | 731
16 | 4        12 | 03       3 | 246     −8 | 7662     −6 | 730
15 | 0        11 | 578      2 | 0368    −9 | 640      −7 | 643
14 | 03       11 | 124      1 | 0456              −8 | 71
13 | 27       10 | 5788     0 | 47                −9 | 3
12 | 269      10 | 24
11 | 1567
10 | 47
```

Box-and-dot plots with levels and spreads removed:

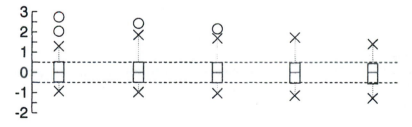

Figure 5.2. Using transformations to correct upward skewness in the batch from Table 5.2.

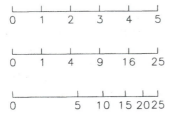

Figure 5.3. Transformation rulers: a "normal" ruler (above); a ruler that would give length measurements with a square transformation (center); the same square transformation ruler with tick marks every 5 units instead of every 1/5 of the length (bottom).

The fifth column in Figure 5.2 shows the results of the $-1/x^2$ transformation. The midspread is now less centered on the median than in the previous transformed batch, but the adjacent values have now reached a more symmetrical distribution. The stem-and-leaf plot shows about as symmetrical a pattern as it is reasonable to expect in a real-world batch of numbers this small. The decision between the last two transformations is difficult. Both have succeeded in eliminating outliers. The adjacent values look more symmetrical in the $-1/x^2$ transformation, while the midspread looks more symmetrical in the $-1/x$ transformation. We might reasonably use the more symmetrical appearance of the stem-and-leaf plot for the $-1/x^2$ transformation to break the tie and opt for the transformed batch in the last column as the most symmetrical, but analysis of either of these batches as a symmetrical shape could be justified.

Transformations often seem an arcane statistical ritual performed more for superstitious reasons than anything else. Their purpose, however, is quite simply to provide a batch of numbers the shape of which makes it possible for the mean and standard deviation to be useful indexes of the center and the spread. The mean and standard deviation are fundamental to many of the techniques discussed further on in this book, and if the mean and standard deviation are not telling us the truth about the center and spread of the batch, then those other techniques will not work well. Transformation can be thought of as measuring with special rulers. Figure 5.3 shows three rulers. At the top is a "normal" ruler that we might use to measure the length of some object. In the middle is a "square transformation ruler." If we measured the same object with this ruler, the result would be the square of the "normal" measurement. The bottom ruler is a square transformation ruler just like the middle ruler, except that the tick marks, instead of being evenly spaced along the ruler, are placed every five units. This shows the way the units of measurement are distributed differently along the square transformation ruler than they are along the "normal" ruler and may provide a better common-sense feel for just how it is that the square transformation shifts numbers along the scale to change the shape of a batch. The same is true of all the other transformations we have discussed. Using them amounts to nothing more than measuring with a peculiar ruler. Although it certainly seems strange at first thought, there is no reason that we couldn't use rulers like the square transformation rulers in Figure 5.3. We would just have to measure everything we wanted to compare with the same peculiar ruler. And indeed, that is exactly what we are doing when we transform a batch of numbers—we are measuring with a peculiar ruler, and we must use the same peculiar ruler (or transformation) on all the batches we want to be able to compare.

THE NORMAL DISTRIBUTION

The single-peaked symmetrical shape we have been pursuing with transformations in this chapter has the essential characteristics of the (in)famous *normal distribution*. Actually, the normal distribution implies some other more specific characteristics, but in practical terms, a batch that is single-peaked and symmetrical can be taken as close enough to a normal distribution to apply statistical techniques suitable for batches that are normally distributed.

The requirement of normality in the distributions of batches that are to be analyzed in certain ways is no deep mathematical mystery. It is not a question of abiding by some secret and sacred principle understandable only to the high priests of statistics. Understanding the importance of the normal distribution begins with such simple and intuitively intelligible notions as the ways in which numerical indexes of level and spread work on asymmetrical batches. We have seen that these numerical indexes simply do not work properly when applied to a batch that is not single-peaked and symmetrical. This is the starting point for understanding why some statistical techniques must only be applied to normal distributions. Many of them begin by characterizing the batch to be analyzed with the mean and standard deviation. If these numerical indexes do not provide accurate and meaningful measures of the center and spread of the batch, then no technique that takes them as a starting point can be expected to produce accurate and meaningful results.

To summarize, then, if we wish to study a batch of numbers that is not single-peaked and symmetrical, we must often take special action. This consists, first, of splitting a batch with multiple peaks into multiple separate batches, each with a single peak. Second, we can use transformations to make the shape of a single-peaked batch more symmetrical. These initial data preparation steps are important to the success of many statistical techniques and must not be overlooked.

Picking the best transformation is a question of subjective judgment. It requires a bit of practice to look at distributions and decide which is most symmetrical. These judgments are especially difficult in small batches of numbers where displacing a single number in the stem-and-leaf plot can have a strong effect on the apparent symmetry. It is a good idea not to be too strongly swayed by appearances that could be changed if only one or two numbers in the batch were slightly different. It is better, instead, to concentrate more heavily on major trends that would only be altered if many numbers were changed.

Picking the best transformation is also a process of trial and error. Although Table 5.3 might help you guess which transformation to try, it is almost always necessary to try several transformations and look at the results (by examining stem-and-leaf and box-and-dot plots of the transformed batches) in order to decide which produces the most symmetrical shape. Compromises are often required, especially when transformations are being applied to two or more batches that eventually are to be compared. The same transformation may not produce the most symmetrical shape for each of the batches involved, but the same transformation must be applied to all the batches if they are to be compared after transformation.

An alternative to correcting asymmetry through transformations is to use statistical techniques based not on the mean and standard deviation but on other indexes—like the trimmed mean and trimmed standard deviation—that are more resistant to the effects of outliers and asymmetry. Such approaches will be discussed where relevant in the following chapters. Generally, if the presence of outliers presents a problem for the use of means and standard deviations, use of the trimmed mean and standard deviation is a good solution. If pervasive asymmetry in the distribution is the problem, applying an appropriate transformation is more effective.

PRACTICE

1. Look carefully at the shapes of the batches of areas of Early and Late Bronze Age sites near Nanxiong from Table 3.5. In the practice questions from Chapters 3 and 4, you have already made stem-and-leaf and box-and-dot plots of these batches. Does either batch have a skewed shape? If so, is it skewed upward or downward?

2. If either Early or Late Bronze Age site areas are skewed, use your statpack to experiment with transformations to correct the asymmetry. Which transformation would you choose for each batch individually? Why? Which transformation would you use for both if you intended to compare the transformed batches? Why?

Chapter 6

Categories

The batches of numbers that we have discussed in Chapters 1–5 have all consisted of a set of measurements of some kind. There is a fundamentally different kind of batch that we must discuss before continuing with the second section of this book. This other kind of batch results from observations of characteristics that are not measured exactly, but instead are grouped into different *categories*. Archaeologists are quite accustomed to the notion of categorizing things. We usually discuss this under the heading of typology, and the definition of typologies (or sets of categories) for artifacts is widely recognized as a fundamental initial step in description and analysis. Much has been written about the "correct" way to define pottery types in particular. Our concern here is not how to define categories, but rather what to do with the result once we have defined the categories and have counted up how many of each there are. When we classify the ceramics from a site as Fidencio Coarse, Atoyac Yellow-White, and Socorro Fine Gray, we are dealing in categories. When we count the number of flakes, blades, bifaces, or debitage from a site, we are also dealing in categories. When we add up the number of cave sites and open sites in a region, we are once again dealing in categories. When we divide the sites in a region into large sites, medium-sized sites, and small sites, we are still once more dealing in categories. Data recorded in terms of such categories compose batches just as do data recorded as true measurements (for example, in centimeters, grams, hectares, etc.).

Table 6.1 provides an example of such categorical information for a set of 140 pottery sherds. One observation made about each sherd is the site where it was recovered. There are three categories here: the Oak Grove site, the Maple Knoll site, and the Cypress Swamp site. A second observation concerns incised decoration, with two categories: each sherd is either incised

Kinds of Data

Some statistics books begin with a chapter about different kinds of data as if the recognition of a standard set of several fundamentally different kinds of data were the rock upon which all statistical analysis was built. The "fundamentally" different kinds, however, are defined in different ways by different authors, and many books don't make much of such distinctions at all. There are almost as many sets of terms as there are authors, and some of the same terms are used in contradictory ways by different authors. The point is that there are a number of characteristics of batches of numbers that vary. You can analyze batches with certain characteristics in one way, but batches that lack those characteristics can be analyzed in a different way. The most important distinction made in this book is between what are here called measurements and categories.

Measurements are things like lengths, widths, areas, weights, and so on— quantities we measure along a scale of appropriate units. True measurements come in fractional values as well as whole number values (difficult numbers to work with like 3-$^{13}/_{16}$ inches, or much easier numbers to work with like 9.68 cm), and there is, in principle at least, an infinite number of potential values along the scale. Measurements, as the term is used in this book, also include numbers of things, like numbers of inhabitants in different regions, numbers of artifacts in different sites, and so on. Measurements can also be derived from other measurements arithmetically. Densities of artifacts in different excavation units, for example, are measurements derived by dividing the numbers of artifacts encountered in each excavation unit by the volume of deposit excavated to arrive at the numbers of artifacts per cubic meter for each excavation unit. In exploratory data analysis, measurements are sometimes called *amounts* (measurements along a scale) and *counts* (counted numbers of things) and *balances* (which can have positive and negative values). Measurements are made along what are sometimes called ratio scales or interval scales. A *ratio scale* has a meaningful zero point, as, for example, a length or weight. An *interval scale* has an arbitrary zero point, which prevents some kinds of manipulations. The usual example of a scale with an arbitrary zero point is temperature. The fact that the zero point is arbitrary means that you really can't say that 60° is twice as hot as 30° (on either Fahrenheit or Celsius scales, although you can do such things with the temperature scale measured from absolute zero).

Categories are essentially groups of things, and we count the numbers of things in each group. We ordinarily work with sets of categories that are *mutually exclusive and exhaustive*. That is, each thing in the set of things we are studying must fit into one category and only one category. Pottery types are a common kind of category in archaeology, and we recognize that pottery types need to be defined so that each sherd can be placed in one and only one type. Colors represent another set of categories. We may sort things out as red, blue, or green. If we find bluish green things, we may need to add a fourth category to the set. Categories are sometimes called *nominal data*. If the categories can logically be arranged in a specific order, then they form *ordinal data*, or *ranks*. Pottery types do not have this property. Categories like large, medium, small, and tiny do; we recognize that to say small, large, tiny, medium is to put these categories out of order. If there are very many categories in the set, ranks begin to act like true measurements in some ways.

The most important distinction between kinds of data for the organization of thoughts in this book is between what we will call measurements and categories, but we will also consider some special treatments that can be applied to ranks—treatments that relate strongly to things we can do with true measurements.

or unincised. This may seem an unusual way to present such information, and indeed it is. The presentation is cumbersome, and it tells us virtually nothing about patterns. In an instance like this, we would much more likely present the information as a *tabulation*, and that is what we will do shortly. It is often convenient to manage categorical information in the manner in which it is presented in Table 6.1, however, especially with computers. Thus it is important to recognize Table 6.1 as one means of organizing the same information we will see in more familiar form in the following tables. Table 6.1 is the most complete and detailed way of recording this information and the most similar to the way in which batches of measurements were initially presented in previous chapters.

Table 6.2 presents the information about where the sherds were recovered in a more compact, familiar, and meaningful way. This simple tabulation of *frequencies* (or *counts*) and *proportions* (or *percentages*) immediately tells us something about how much pottery came from where—something that was not at all apparent in Table 6.1. More pottery came from Oak Grove than any other site, and the least came from Maple Knoll. Table 6.3 performs the same task for the information about decoration for the same 140 sherds. Most of the sherds are unincised, but the difference in proportions between incised and unincised is not extreme.

In effect, Table 6.1 contains sets of related batches, like the related batches that have been discussed in previous chapters. In this case, we could divide the sherds into three related batches, as in Table 6.2: sherds from the Oak Grove site, sherds from the Maple Knoll site, and sherds from the Cypress Swamp site. Or we could divide them in a different way into two related batches, as in Table 6.3: incised sherds and unincised sherds. Each set of categories is simply one way of dividing the whole set of sherds into different batches. We might well want to compare the first three batches (the

Table 6.1. Information about 140 Pottery Sherds

Site where found	Decoration	Site where found	Decoration	Site where found	Decoration
Oak Grove	Unincised	Maple Knoll	Unincised	Cypress Swamp	Incised
Maple Knoll	Incised	Oak Grove	Incised	Cypress Swamp	Unincised
Cypress Swamp	Unincised	Oak Grove	Unincised	Cypress Swamp	Unincised
Cypress Swamp	Incised	Oak Grove	Unincised	Oak Grove	Incised
Cypress Swamp	Incised	Maple Knoll	Incised	Oak Grove	Unincised
Cypress Swamp	Unincised	Cypress Swamp	Unincised	Maple Knoll	Incised
Cypress Swamp	Incised	Cypress Swamp	Unincised	Maple Knoll	Unincised
Oak Grove	Incised	Oak Grove	Incised	Oak Grove	Incised
Oak Grove	Unincised	Oak Grove	Unincised	Oak Grove	Unincised
Maple Knoll	Unincised	Maple Knoll	Unincised	Maple Knoll	Incised
Oak Grove	Incised	Cypress Swamp	Incised	Cypress Swamp	Incised
Oak Grove	Unincised	Cypress Swamp	Incised	Cypress Swamp	Unincised
Maple Knoll	Incised	Oak Grove	Incised	Oak Grove	Incised
Maple Knoll	Unincised	Oak Grove	Unincised	Oak Grove	Unincised
Cypress Swamp	Unincised	Maple Knoll	Unincised	Oak Grove	Unincised
Cypress Swamp	Incised	Cypress Swamp	Unincised	Maple Knoll	Incised
Oak Grove	Unincised	Cypress Swamp	Incised	Cypress Swamp	Unincised
Maple Knoll	Incised	Oak Grove	Incised	Cypress Swamp	Incised
Maple Knoll	Unincised	Oak Grove	Unincised	Oak Grove	Incised
Cypress Swamp	Incised	Maple Knoll	Unincised	Oak Grove	Unincised
Oak Grove	Incised	Cypress Swamp	Unincised	Maple Knoll	Unincised
Oak Grove	Unincised	Cypress Swamp	Incised	Maple Knoll	Incised
Maple Knoll	Unincised	Oak Grove	Unincised	Oak Grove	Incised
Cypress Swamp	Unincised	Maple Knoll	Incised	Oak Grove	Unincised
Oak Grove	Incised	Cypress Swamp	Unincised	Cypress Swamp	Incised
Oak Grove	Unincised	Oak Grove	Unincised	Cypress Swamp	Unincised
Maple Knoll	Incised	Maple Knoll	Incised	Cypress Swamp	Unincised
Maple Knoll	Unincised	Cypress Swamp	Unincised	Oak Grove	Incised
Oak Grove	Incised	Cypress Swamp	Incised	Oak Grove	Unincised
Oak Grove	Unincised	Oak Grove	Incised	Maple Knoll	Unincised
Maple Knoll	Incised	Oak Grove	Unincised	Cypress Swamp	Incised
Cypress Swamp	Incised	Oak Grove	Unincised	Oak Grove	Incised
Cypress Swamp	Unincised	Maple Knoll	Incised	Oak Grove	Unincised
Oak Grove	Incised	Cypress Swamp	Unincised	Oak Grove	Unincised
Oak Grove	Unincised	Oak Grove	Incised	Maple Knoll	Incised
Maple Knoll	Incised	Oak Grove	Unincised	Cypress Swamp	Unincised
Maple Knoll	Unincised	Oak Grove	Unincised	Oak Grove	Incised
Cypress Swamp	Unincised	Maple Knoll	Incised	Oak Grove	Unincised
Maple Knoll	Incised	Cypress Swamp	Unincised	Maple Knoll	Unincised
Cypress Swamp	Unincised	Oak Grove	Incised	Cypress Swamp	Unincised
Oak Grove	Incised	Oak Grove	Unincised	Cypress Swamp	Incised
Oak Grove	Unincised	Maple Knoll	Incised	Cypress Swamp	Unincised
Maple Knoll	Unincised	Cypress Swamp	Incised	Oak Grove	Incised
Cypress Swamp	Unincised	Cypress Swamp	Unincised	Oak Grove	Unincised
Maple Knoll	Incised	Oak Grove	Incised	Maple Knoll	Incised
Oak Grove	Incised	Oak Grove	Unincised	Maple Knoll	Unincised
Oak Grove	Unincised	Maple Knoll	Incised		

Table 6.2. Sherds from Three Sites

	Oak Grove	Maple Knoll	Cypress Swamp	Total
Frequency	59	37	44	140
Proportion	42.1%	26.4%	31.4%	99.9%

Table 6.3. Pottery Decoration

	Incised	Unincised	Total
Frequency	64	76	140
Proportion	45.7%	54.3%	100.0%

sherds from each of the three sites) in regard to the other set of categories (incised decoration). Table 6.4 extends the tabulations of Tables 6.2 and 6.3 to accomplish this comparative goal by simultaneously dividing the sherds by site and by incised decoration. Such a tabulation is sometimes called a *two-way table*, because it divides the entire set of sherds into categories in two different ways simultaneously. In this kind of table there are also two different ways to use percentages in comparing these batches.

COLUMN AND ROW PROPORTIONS

Following the frequencies in Table 6.4a are *column proportions* (Table 6.4b). These proportions are similar to those in Table 6.3, but now they are calculated separately for each of the three sites. The average column proportions at the extreme right of Table 6.4b are not simply the averages of the

Table 6.4. Incised and Unincised Sherds from Three Sites

a. Frequencies

	Oak Grove	Maple Knoll	Cypress Swamp	Total
Incised	25	21	18	64
Unincised	34	16	26	76
Total	59	37	44	140

b. Column proportions

	Oak Grove	Maple Knoll	Cypress Swamp	Average
Incised	42.4%	56.8%	40.9%	45.7%
Unincised	57.6%	43.2%	59.1%	54.3%
Total	100.0%	100.0%	100.0%	100.0%

c. Row proportions

	Oak Grove	Maple Knoll	Cypress Swamp	Total
Incised	39.1%	32.8%	28.1%	100.0%
Unincised	44.7%	21.1%	34.2%	100.0%
Average	42.1%	26.4%	31.4%	99.9%

Calculating Percentages and Rounding Error

We have noted *rounding error* before, but Tables 6.2 and 6.3 provide an opportunity to clear up this little mystery completely. We know that 140 sherds are
100%, and the percentages in Table 6.3 add up to 100.0, but the percentages of
the three categories in Table 6.2 add up to only 99.9%. In both tables the
percentages have been rounded off to one digit following the decimal point.
For Table 6.3, the full calculations of the percentages are 64/140 =
.4571428571428571428... and 76/140 = .5428571428571428571...

Both of these numbers will continue to repeat the same sequence of
digits (... 142857...) forever. The division will never come out even, no matter
how far out it is carried. To change .4571428571428571428... and
.5428571428571428571... from ordinary decimal fractions into percentages,
of course, we multiply them by 100: 45.71428571428571428...% and
54.28571428571428571...% [And while we're on the subject, it is worth emphasizing that 0.45 and 45.0% are the same number. There is a big difference
between 0.45 and 0.45%—0.45 means .45 out of 1.00 or 45 out of 100 or
4,500 out of 10,000 (that is, almost half), but 0.45% means 0.45 out of 100 or
45 out of 10,000 (or far less than half). It is essential to be careful with a
decimal point and the % symbol.] Clearly the percentages we have here must
be rounded off. If we want one digit after the decimal place in the percentage,
45.71428571428571428...% rounds *down* to 45.7% and we lose the extra
.01428571428571428... %, and 54.28571428571428571...% rounds *up* to
54.3%. In rounding this number up we have actually added
.01428571428571428... to it, which is exactly the amount that we lost in
rounding the other percentage down. Since one percentage has been raised by
the same amount that the other has been lowered by, the amounts cancel each
other out when we add the percentages together, and the total is 100.0%.

In the percentages in Table 6.2, however, all three turn out to round
down:

- 59/140 = .42142857142857... which rounds down to .421 (or 42.1%),
  losing .042857142857...%.
- 37/140 = .26428571428571... which rounds down to .264 (or 26.4%),
  losing .028571428571...%.
- 44/140 = .31428571428571... which rounds down to .314 or 31.4%,
  losing .028571428571...%.

If we add up what we have lost in rounding all three percentages down,
we get almost exactly the 0.1% that is missing from the 99.9% total of the
three rounded off percentages:

(.042857142857...% + .028571428571...% + .028571428571...%
= .099999999999...%)

Precisely the same thing can happen in the other direction if more is
gained by rounding some percentages up than is lost by rounding others

down. Thus the total of a set of percentages can be slightly more than 100%. Sometimes, doing percentage calculations with more decimal digits of precision will remove rounding error. In a case like this example, however, where the quotient of the division repeats infinitely, it doesn't matter how far out we carry the calculations. They will never come out *exactly* even. Sooner or later we have to round off and accept a little rounding error.

individual site proportions. They are actually the same as the proportions in Table 6.3. That is, they are the proportions for the complete set of sherds considering all sites together.

Column proportions are useful for comparing columns to each other—in this instance comparing the three sites to each other with regard to their relative proportions of incised and unincised sherds. The average column proportions provide a standard against which to compare each site. With 56.8% incised sherds, the Maple Knoll site is well above the average of 45.7%, while the Oak Grove site falls slightly below average, and the Cypress Swamp site falls farther below average.

The bar graph in Figure 6.1, a relative of the histogram, provides a familiar way to show these patterns graphically. The four bars at the left illustrate the proportions of incised sherds at the three sites and how those proportions compare to the average proportion of incised sherds. The bar representing the average is shaded in order to set it off from the others and call attention to it as the standard against which the others are compared. Many computer programs draw bar graphs, and most will dress them up in very handsome ways with patterns of shading, colors, perspective, fake shadows, and so on. These elaborations can be very decorative, but it is also easy to get carried away with them and produce a graph that is actually more difficult to read than a very simple one. Circular *pie charts*, another means of illustrating percentages, are in such common usage that it is really not necessary to discuss them here. Bar graphs facilitate more precise comparisons and make it possible to include the average proportion as a standard of comparison.

Looking at the bar graph in Figure 6.1, it is clear that the Maple Knoll site stands out for a well above average proportion of incised sherds. Oak Grove and Cypress Swamp both have slightly lower than average proportions of incised sherds, but their departures from average are not as strong as the departure from average at the Maple Knoll site. We see, of course, exactly the opposite picture when we look at the bars to the right in Figure 6.1, representing proportions of unincised sherds. Maple Knoll is substantially below average in proportion of unincised sherds, while Oak Grove and Cypress Swamp are both slightly above average in their proportions of unincised sherds.

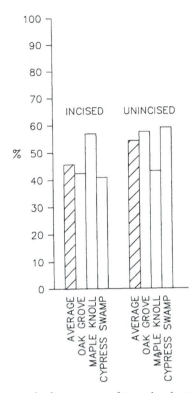

Figure 6.1. Bar graph of proportions of incised and unincised sherds.

Table 6.4c provides *row proportions* for the table. The row proportions relate to those in Table 6.2, but now they are calculated separately for incised and unincised sherds. The average row proportions at the bottom of Table 6.4c are, once again, not the averages of the individual row proportions, but the proportions for the entire set of sherds taken together, combining incised and unincised sherds. They are exactly the same as the proportions from Table 6.2. Row proportions are not very helpful for comparing one column to another, but they are quite useful for comparing one row to another. Once again, the average proportions provide a standard for comparison—this time against which to compare the individual rows. With 32.8%, incised sherds are more common than average at the Maple Knoll site since only 26.4% of the sherds overall come from Maple Knoll. Correspondingly, incised sherds are less common than average at the Oak Grove and Cypress Swamp sites.

The bar graph in Figure 6.2 provides a graphical comparison of rows based on the row percentages in Table 6.4c. Incised sherds are less common than average at Oak Grove and at Cypress Swamp, while unincised sherds are

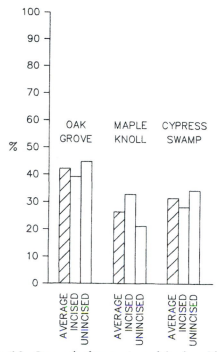

Figure 6.2. Bar graph of proportions of sherds at different sites.

more common than average. At Maple Knoll departures from average are stronger than at either of the other two sites, and in the opposite direction, with incised sherds being more common.

Categories enable us to break a batch down into subbatches that can then be compared to each other. The comparison may be of another set of categories, as in the example in this chapter, or it may be of a measurement. If, for example, we measured the approximate diameter of the vessel represented by each of the sherds from Table 6.1, we could then break the batch of sherds into subbatches according to site and compare vessel diameters for the three sites. The tools needed for that comparison are precisely those already discussed in Chapters 1–4. We could, for example, draw three box-and-dot plots (one for the subbatch representing each site) all at the same scale to compare rim diameters of vessels at the three sites.

PRACTICE

1. Beginning to assess settlement distribution in the area around Al-Amadiyah, you select 400 random points on the map of your study

Be Careful How You Say It

It is extremely easy to get confused about row and column proportions and even easier to talk about them in ways that confuse whoever you're talking to. We can, for example, look at Table 6.4c and notice that the 39.1% for incised sherds from the Oak Grove site is larger than the percentage for incised sherds from either of the other two sites. It would be very easy, then, to go on to say, "Oak Grove has the highest percentage of incised sherds (39.1% versus 32.8% and 28.1% for the other two sites)." This statement is incomplete, however, in a critical way: it doesn't say percent *of what*. The percentages referred to are percentage *of incised sherds*. Thus, the sentence really is telling us that 39.1% *of the incised sherds found* come from Oak Grove—more incised sherds than from any other site. It would be extremely easy for the reader of that sentence, however, to take it to mean that 39.1% *of the sherds from Oak Grove* are incised, compared to smaller percentages for other sites. (Go back and read the sentence again—it really could mean either of these two quite different things.) Actually 42.4% of the sherds from Oak Grove are incised, and an even higher percentage of the sherds from the Maple Knoll site are incised. It is only the Maple Knoll site that provided a collection in which more than half the sherds were incised.

If you want to compare sites in this way according to the proportional composition of their ceramic assemblages, you must calculate proportions of different categories as a percentage of the site assemblage rather than as a percentage of the total number of that category found at all sites. Saying that 39.1% of the incised sherds came from Oak Grove, more than from any other site, and implying that incised ceramics were more "popular" there than elsewhere, is untrue. A higher percentage of the unincised sherds came from Oak Grove, too. For an archaeologist, this is a fairly trivial observation; we simply have a larger collection of sherds from Oak Grove than from anywhere else, for reasons that are probably entirely uninteresting, so it is no surprise that we have more of all kinds of sherds from there than from anywhere else. Using average proportions, as discussed in this chapter, as a standard against which to compare the proportions of different categories can help to avoid this confusion and help to make clear of what the percentages are.

area and visit each one in the field. You classify each of the 400 points according to its setting (alluvial valley floor, rocky piedmont, or steeper mountain slopes), and you observe whether there is any evidence of prehistoric occupation there. Your results are as follows:

- 41 points are in the alluvial valley floor. Of these, 14 show evidence of prehistoric occupation, and 27 do not.

Table 6.5. Areas of Sites in Three Environmental Settings in the Study Area at Al-Amadiyah

Site area (ha)	Setting	Site area (ha)	Setting	Site area (ha)	Setting
2.8	Piedmont	8.5	Piedmont	2.3	Piedmont
2.3	Piedmont	2.3	Piedmont	1.1	Piedmont
4.2	Piedmont	3.4	Piedmont	0.9	Slopes
19.0	Alluvium	4.9	Alluvium	3.6	Piedmont
5.3	Piedmont	2.5	Piedmont	2.1	Piedmont
7.7	Alluvium	5.5	Piedmont	2.0	Slopes
15.8	Alluvium	4.1	Piedmont	2.2	Piedmont
1.1	Slopes	5.8	Piedmont	2.5	Piedmont
6.7	Alluvium	1.0	Slopes	1.0	Slopes
10.5	Alluvium	2.9	Piedmont	1.3	Slopes
9.3	Alluvium	6.9	Piedmont	2.3	Piedmont
7.9	Piedmont	7.4	Piedmont	2.7	Piedmont
3.2	Piedmont	0.5	Slopes	2.9	Slopes
20.3	Alluvium	4.5	Piedmont	0.8	Slopes
6.4	Piedmont	5.2	Piedmont	4.2	Piedmont
7.7	Piedmont	2.0	Piedmont	1.4	Slopes
4.9	Piedmont	0.3	Slopes	1.2	Slopes
9.3	Alluvium	3.9	Piedmont	5.3	Piedmont
1.9	Piedmont	3.0	Piedmont	10.2	Alluvium
6.2	Alluvium	3.5	Piedmont	8.8	Piedmont
7.3	Piedmont	4.5	Piedmont	1.3	Slopes
3.1	Piedmont	3.3	Piedmont	8.8	Alluvium
5.7	Piedmont	1.5	Piedmont	8.3	Piedmont
2.4	Piedmont	2.1	Piedmont	0.7	Slopes
5.3	Piedmont	4.2	Piedmont	4.8	Piedmont
7.2	Piedmont	3.0	Piedmont	1.5	Slopes
6.7	Piedmont	4.1	Piedmont	0.8	Slopes
0.4	Slopes	7.7	Alluvium	9.8	Piedmont
6.0	Piedmont	3.7	Piedmont	3.2	Piedmont
4.0	Piedmont	2.0	Piedmont	0.7	Slopes
2.6	Piedmont	3.5	Piedmont	17.7	Alluvium
4.7	Piedmont	0.8	Slopes	0.5	Slopes
8.1	Piedmont	2.9	Piedmont		

- 216 points are in the stony piedmont. Of these, 64 show evidence of prehistoric occupation, and 152 do not.
- 143 points are on the steeper mountain slopes. Of these, 20 show evidence of prehistoric occupation, and 123 do not.
- Use proportions and a bar chart to compare the three environmental settings in regard to the density of prehistoric occupation your preliminary fieldwork found in each. Would you say that some zones were more intensively occupied than others by prehistoric inhabitants? If so, which one(s)?

2. You continue your study at Al-Amadiyah by revisiting each of the 98 locations that did show evidence of prehistoric occupation, and you measure the areal extent of the surface scatters of artifacts that indicate the archeological sites. Your results are presented in Table 6.5. Use the environmental settings to separate the 98 locations into three separate batches, and use box-and-dot plots to compare the three batches in regard to site area. Do site sizes appear to differ from one setting to another? Just how?

Part II
Random Sampling

Chapter *7*

Samples and Populations

The notion of sampling is at the very heart of the statistical principles discussed in this book, so it is worth pausing here at the beginning of Part II to discuss clearly what sampling is and to consider some of the issues that the practice of sampling raises in archaeology. Archaeologists have, in fact, been practicing sampling in one way or another ever since there were archaeologists, but widespread recognition of this fact has only come about in the past 20 years or so. In 1970 the entire literature on sampling in archaeology consisted of a very small handful of chapters and articles. Today there are hundreds and hundreds of articles, chapters, and whole books, including many that attempt to explain the basics of statistical sampling to archaeologists who do not understand sampling principles.

Unfortunately, many of these articles seem to have been written by archaeologists who do not themselves understand the most basic principles of sampling. The result has been a great deal of confusion. It is possible to find in print (in otherwise respectable journals and books) the most remarkable range of contradictory advice on sampling in archaeology, all supposedly based on clear statistical principles. At one extreme is the advice that taking a 5% sample is a good rule of thumb for general practice. At the other extreme is the advice that sampling is of no utility in archaeology at all because it is impossible to get any relevant information from a sample or because the materials archaeologists work with are always incomplete collections anyway and one cannot sample from a sample. (For reasons that I hope will be clear well before the end of this book, both these pieces of advice are wrong.) It turns out that good sampling practice requires not the memorization of a series of arcane rules and procedures but rather the understanding of a few simple principles and the thoughtful application of considerable quantities of common sense.

There is another way, too, in which a pause for careful consideration of sampling principles can be useful to archaeologists. A lament about the regrettably small size or the questionable representativeness of the sample is a common conclusion to archaeological reports. Statisticians have put a great deal of energy into thinking about how we can work with samples. Some of the specific tools they have developed could be used to considerably more advantage in archaeology than they often have been in the past, and this book attempts to introduce several of them. More fundamentally, though, at least some of the logic of working with samples on which statistical techniques are based is equally relevant to other (nonstatistical) ways of making conclusions from samples. Clear thinking about the statistical use of samples can pay off by helping us better understand other kinds of things we might do with samples as well.

WHAT IS SAMPLING?

Sampling is the selection of a sample of elements from a larger *population* (sometimes called a *universe*) of elements for the purpose of making certain kinds of inferences about that larger population as a whole. The larger population, then, consists of the set of things we want to know about. This population could consist of all the archaeological sites in a region, all the house floors pertaining to a particular period, all the projectile points of a certain archaeological culture, all the debitage in a specific midden deposit, etc. In these four examples, the elements to be studied are sites, house floors, projectile points, and debitage, respectively. In order to learn about any of these populations, we might select a smaller sample of the elements of which they are composed. The key is that we wish to find something out about an entire population by studying only a sample from it.

WHY SAMPLE?

It at first seems to make sense to say that the best way to find out about a population of elements is to study the whole population. Whenever one makes inferences about a population on the basis of a sample there is some risk of error. Indeed, sampling is often treated as a second-best solution in this regard—we sample when we simply cannot study the entire population. Archaeologists are almost always in precisely this situation. If the population we are interested in consists of all the sites in a region, almost certainly some of the sites have been completely and irrevocably destroyed by more recent human activities or natural processes. This problem occurs not only at the regional level—rare is the site where none of the earlier deposits have been destroyed or damaged by subsequent events. In this typical archaeological situation the entire population we might wish to study is simply not available for study. We are forced to make inferences about it on the basis of a smaller sample, and it does no good to simply close our eyes and insist otherwise. The unavailability of entire populations for study raises some particularly vexing issues in archaeology, to which we will return below.

We might not be able to study even the entire available population because it would be prohibitively expensive, because it would take too much time, or for other reasons. One of the most interesting other reasons is that studying an element may destroy it. It might be interesting to contemplate submitting an entire population of, say, prehistoric corncobs for radiocarbon dating, but we are unlikely to do so since afterward there would be no corncobs for future study of other kinds. We might choose to date a sample of the corncobs, however, in order to make inferences about the age of the population while reserving most of its elements for other sorts of study.

In cases where destructiveness of testing or limitations of resources, time, or availability interfere with our ability to study an entire population, it is fair to say that we are forced to sample. Precisely such conditions often apply in the real world, so it is common for archaeologists to approach sampling somewhat wistfully—wishing they could study the entire population but grudgingly accepting the inevitability of working with a sample. Perhaps the most common situation in which such a decision is familiar concerns determination of the sources of raw materials for the manufacture of ceramics or lithics. At least some techniques for making such identifications are well established, but they tend to be time-consuming, costly, and/or destructive. So, while wishing to know the raw material sources for an entire population of artifacts, we often accept such knowledge for only a sample from the population.

Often, however, far from being forced to sample, we should choose to sample because we can find out more about a population from a sample than by studying the entire population. This paradox arises from the fact that samples can frequently be studied with considerably greater care and precision than entire populations can. The gain in knowledge from such careful study of a sample may far outweigh the risk of error in making inferences about the population based on a sample. This principle is widely recognized, for example, in census-taking. Substantial errors are routinely recognized in censuses, resulting at least in part from the sheer magnitude of the counting task. When a population consists of millions and millions of elements it is simply not possible to treat the study of each element in the population with the same care taken with each element in a much smaller sample. As a consequence, national censuses regularly attempt to collect only minimal information about the entire population, and much more detailed information about a much smaller sample. It is even becoming increasingly common for the minimal information collected for the entire population to be "corrected" on the basis of a more careful study of a smaller sample!

Archaeologists are frequently in a similar position. Certain artifact or ecofact categories from even a modest-sized excavation may well number far into the thousands. Detailed characterization of lithic debitage, for example, can be a very time-consuming process. We are likely to learn considerably more from a detailed study of a sample of lithic debitage than from a cursory study of an entire population of thousands of waste flakes. In such situations we should eagerly embrace effective sampling techniques as an improvement over the study of entire populations.

HOW DO WE SAMPLE?

If the purpose of sampling is to make inferences about a population on the basis of a smaller sample of elements selected from it, then it is important to select the sample in such a way as to maximize the chance that it accurately represents the population from which it is selected. *Random sampling* is a very effective way to maximize this chance of accuracy, and we select samples randomly whenever possible.

We are all familiar with many ways to select samples randomly. The practices of drawing straws, drawing names from a hat, and turning containers round and round to spill out bingo numbers are all efforts at random selection. Such physical methods seldom achieve true randomness, but a proliferation of governmental lotteries has spawned a multitude of mechanical contraptions that select at least very nearly truly random numbers (all in a manner designed to be engaging to the home audience watching the drawing on television).

Perhaps the most common means of selecting a random sample is to number each element in the population from which the sample is to be drawn (from 1 to however many there are in the population). A list of random numbers then identifies the elements that will make up the sample. The list of random numbers may come from a computer program or it may come from a random number table such as Table 7.1.

Suppose you want a list of 10 random two-digit numbers (that is, numbers between 00 and 99). Pick a number in the table to use as a starting

Table 7.1. Random Numbers

```
50 79 13  18 85 26  80 01 74  73 44 03  81 25 58  14 74 59  91 56 48  88 67 99  04 91 80
17 97 55  39 91 18  43 28 73  68 74 25  62 87 14  53 69 21  35 22 37  12 45 85  14 74 75
38 48 77  82 81 82  47 75 62  63 44 62  38 12 64  22 93 81  52 10 62  45 07 53  74 39 93
76 87 58  73 88 35  35 16 46  31 38 60  51 36 31  55 34 69  09 34 67  60 31 73  10 37 43
51 50 51  63 43 65  96 06 63  89 93 36  02 25 02  47 75 46  02 50 01  72 55 10  56 69 09

96 65 34  00 41 60  29 64 23  61 71 94  61 38 48  70 10 91  48 83 73  02 93 32  08 69 07
91 22 76  00 63 04  07 14 17  18 60 19  11 75 72  86 97 67  69 98 09  11 98 17  52 99 69
28 99 59  78 92 33  29 54 62  17 78 29  57 52 54  74 64 14  20 47 00  94 97 43  46 33 07
81 53 42  15 05 38  14 09 83  44 66 04  06 10 42  14 28 62  75 62 28  49 00 75  52 48 09
32 95 82  45 22 67  42 78 47  47 19 89  18 84 62  24 49 82  40 00 97  99 13 75  46 75 18

59 25 27  06 30 60  19 87 34  27 10 04  94 28 21  59 82 96  16 68 69  74 36 58  19 90 19
01 41 23  34 37 75  30 24 21  41 34 04  18 18 74  66 91 46  27 09 99  91 20 19  33 59 60
34 58 27  03 62 01  58 59 98  01 86 10  12 08 74  52 23 66  42 85 72  02 49 45  22 60 68
61 33 38  19 16 16  71 71 61  23 70 21  57 63 95  14 91 04  47 37 98  26 77 37  95 34 20
91 75 95  57 13 78  90 20 21  42 56 54  36 71 43  42 17 99  06 54 58  81 33 64  92 26 61

40 66 19  64 53 15  27 39 11  28 71 36  65 70 23  34 43 27  89 67 31  31 12 85  80 73 35
80 55 13  01 99 94  72 29 87  73 06 68  87 97 33  27 62 51  52 33 17  72 90 06  72 37 11
45 87 71  15 94 31  09 98 88  64 20 05  11 84 10  14 91 15  80 68 26  56 03 22  10 08 18
19 30 96  02 25 42  68 26 34  79 50 41  64 32 71  90 43 20  91 68 04  07 38 05  30 34 26
60 38 33  50 59 24  73 82 64  65 28 09  32 04 76  63 81 96  83 68 90  52 43 68  89 44 57

22 94 75  27 41 32  86 21 91  49 13 71  57 56 28  12 40 56  03 54 54  47 92 27  29 18 91
25 23 23  20 26 36  48 13 17  54 42 97  63 86 42  64 65 01  69 49 32  87 79 24  49 96 79
59 51 80  91 35 81  29 17 19  19 71 29  76 87 03  97 67 52  21 47 29  20 01 39  33 37 45
05 40 65  66 23 54  23 94 43  44 09 08  81 12 79  58 01 74  81 60 89  70 89 43  37 53 90
61 99 79  13 20 09  56 58 07  59 70 46  32 86 47  36 81 20  89 89 98  71 94 37  88 72 58

24 34 19  08 05 18  51 49 14  30 48 09  47 94 63  12 04 80  76 38 53  09 37 03  04 06 53
29 48 01  18 37 83  94 16 20  37 09 53  63 72 89  96 74 35  13 21 80  77 54 24  09 72 15
65 78 94  61 74 72  11 71 52  15 71 62  98 87 73  39 41 82  12 98 31  83 67 01  86 03 52
04 24 77  46 63 39  03 10 85  10 79 39  08 17 74  64 84 20  43 21 22  46 26 73  51 41 17
73 71 88  69 64 06  08 26 63  51 35 45  66 52 78  38 85 11  80 39 30  86 85 48  44 46 43

88 59 20  63 92 58  52 12 02  37 13 31  42 52 34  77 50 18  09 17 48  46 41 32  83 26 01
84 82 52  27 55 25  20 16 11  66 94 25  04 94 55  79 03 65  61 21 49  97 72 46  56 26 52
82 26 26  52 50 21  63 86 14  11 69 21  98 97 03  68 59 09  98 34 50  58 38 79  03 64 69
81 52 82  82 86 08  45 99 54  14 71 46  14 01 68  33 59 29  71 09 23  37 84 04  92 61 34
90 95 02  61 36 94  98 81 54  90 60 64  84 49 23  92 30 99  69 65 65  47 54 73  17 81 21

37 78 13  13 55 40  07 53 92  98 82 64  01 11 08  94 91 84  83 55 46  30 96 74  13 54 30
01 87 88  82 01 76  59 28 87  03 73 69  22 99 27  30 62 73  02 34 82  30 59 37  27 95 50
02 96 02  54 62 25  36 56 61  38 80 15  93 30 11  34 67 53  81 83 54  83 86 47  64 43 03
40 53 25  64 31 38  89 14 23  54 33 86  58 03 94  57 03 68  78 38 14  20 09 42  82 84 06
46 81 46  18 47 75  70 20 70  33 15 43  73 67 61  05 55 50  03 15 86  55 91 52  73 90 95
```

(continued)

Table 7.1. (*Continued*)

69 72 68	17 87 22	62 08 49	40 32 38	25 71 59	29 67 81	23 68 36	49 94 65	15 03 72
26 24 90	53 49 35	91 07 60	74 61 62	06 07 67	95 99 56	28 56 02	52 61 94	81 14 33
68 17 38	10 48 60	81 73 25	34 55 76	40 84 05	23 55 96	20 60 74	08 03 42	51 81 07
06 51 06	07 44 30	86 12 69	99 16 51	10 05 54	16 07 18	16 24 26	09 97 30	57 50 11
45 52 21	16 03 36	28 32 27	25 44 46	14 17 81	29 86 97	59 12 03	67 28 83	33 03 64
54 72 12	20 91 87	53 87 29	39 84 26	59 80 66	44 84 84	63 77 81	31 48 92	45 99 33
72 65 08	37 37 55	91 23 02	22 51 88	94 32 45	09 14 81	31 14 27	26 61 93	41 52 08
47 20 65	40 51 39	78 88 88	71 45 86	03 08 99	61 16 56	47 08 54	89 79 29	24 91 42
94 79 42	62 56 17	34 45 56	84 96 09	56 22 13	14 87 21	97 66 60	48 64 56	41 45 92
40 03 28	30 16 77	79 10 05	94 90 35	08 03 11	91 56 83	42 23 20	08 44 82	13 47 70

point by closing your eyes and stabbing your finger at the table. Suppose your finger landed on the number 51 that appears in the third column of the fifth row. You could read off the next 10 numbers across the fifth row so that your 10 random numbers would be 63, 43, 65, 96, 06, 63, 89, 93, 36, and 02. Or you could read down the third column for 34, 76, 59, 42, 82, 27, 23, 27, 38, and 95. Or you could read to the left on the fifth row, for 50 and 51, and then drop down to the sixth row and read back across toward the right to continue with 96, 65, 34, 00, 41, 60, 29, and 64. You can read in either direction column-wise or row-wise from the starting point you select.

The principal rule for proper use of a random number table is never to use it in exactly the same way twice. If you need a second sequence of random numbers, close your eyes again and pick a new starting point, or read in a different direction than you did the previous time (or both). You just do not want to use the same sequence of random numbers over and over again; make a fresh start each time you select a sample. (And watch out for calculators that claim to generate random numbers. Some of them simply generate the same sequence of random numbers every time you press the button.)

If you need one-digit numbers, you can treat each individual column of digits as a separate column in the table. If you need four-digit numbers, you can treat pairs of columns together as four-digit numbers. If you need three-digit numbers you can simply ignore the first or last digit in a four-digit number. The spaces dividing the numbers into columns of two-digit numbers, in short, are entirely arbitrary, as are the wider spaces grouping the columns by threes. Likewise, the extra space setting off groups of five rows each is included only to make the table easier to read.

Suppose the population from which you wish to select a sample contains 536 elements, numbered from 001 to 536. You need a list of three-digit random numbers between 001 and 536. You can select a list of numbers in exactly the manner just described, except that you ignore any number less than 001 (that is, 000) or any number greater than 536 (that is, 537 to 999). You simply skip past these inapplicable numbers in the list and continue to select those in the relevant range until you have as many as needed.

Sometimes the same number will appear more than once in a list of random numbers. If this happens, you can follow either of two courses. The first is to ignore multiple appearances of the same number and continue reading the random number table until you have as many numbers as you need without repetitions. This is called *sampling without replacement*. (The name makes sense if you imagine that you were actually drawing numbered slips from a hat without replacing the slips in the hat for potential reselection on subsequent drawings.) Sampling without replacement is the course of action that seems to make intuitive good sense to many people.

Sampling with replacement, however, turns out to be a little simpler mathematically, and the equations in this book are those for sampling with replacement. In sampling with replacement, each time you draw a numbered slip from the hat (speaking metaphorically), you write down the number and replace the slip in the hat so that it could be drawn again in the future. The analogous procedure, when sampling with a random number table, is to include repeated numbers in the sample as many times as they appear in the list from the random number table. The data for the corresponding elements, then, are included among the sample data as if each occurrence in the sample were an entirely different element.

Suppose, for example, that we were sampling with replacement from a population of scrapers, in an effort to estimate the mean length of scrapers in the population. The random numbers chosen from the table might be 23, 42, 13, 23, and 06. We would select the scrapers with the numbers 06, 13, 23, and 42 and measure the length of each. We would, however, write down five length measurements, not four, so as to include the length measurement for scraper number 23 twice. The number of elements in the sample would also be five, not four. To reemphasize, *it is this procedure, sampling with replacement, that the equations in this book are appropriate for.* Slightly different equations are technically necessary for sampling without replacement, although in almost every practical instance it makes very little meaningful difference in the results. It is, however, quite easy to adhere strictly to the assumptions on which the formulas given in this book are based simply by including the data from an element in the sample as many times as that element is selected.

REPRESENTATIVENESS

The kind of sample selection we have just discussed is called *simple random sampling*. The effect of using a table of random numbers is to give each individual element in the population an equal chance of selection, and this is the most straightforward way to phrase the essential principle of simple random sampling. It is because each individual element in the population has an equal chance of inclusion in the sample that a random sample provides us with our best chance of obtaining a sample that accurately represents the population.

The concept of *representativeness* is a slippery one, worth discussing more fully. As noted at the beginning of this chapter, our aim in sampling is to make inferences about a population of elements based on a sample from that population. We take that sample to represent the population, so the representativeness of the sample is of critical importance. The problem is that, without studying the entire population, we can never be absolutely certain that the sample represents it accurately. If we intend to study the entire population, of course, there is no need to worry about the representativeness of a sample. It is only if we do not intend to study the entire population that we must worry about the representativeness of a sample, but that is precisely the situation in which we cannot provide any guarantee of representativeness. It is this difficulty that much of the rest of this book is about. (Statistics books have more in common with Joseph Heller's *Catch 22* than is often noticed.)

Some archaeologists seem to have the impression that, if a sample has been selected randomly, then it is guaranteed to be representative. Nothing could be farther from the truth. Like samples not selected randomly, random samples represent the populations from which they were selected sometimes quite accurately, sometimes with moderate accuracy, and sometimes very inaccurately. Random sampling, while it does not provide a guarantee, however, does give us our best chance at a representative sample. *Most important of all, random sampling provides a basis for estimating how likely it is that our inferences about the population are wrong, and thus tells us how much confidence we should place in these inferences.*

DIFFERENT KINDS OF SAMPLING AND BIAS

Simple random sampling is, as its name implies, the simplest and most straightforward method of selecting a random sample. In most of the rest of this book, mention of random samples refers to simple random samples. There are other somewhat more complicated variants of random sampling. They are best dealt with after the implications of simple random sampling are fully explored and understood, so we will not discuss them in any detail here. It is important to recognize their existence at this point, though, in order to understand the limits of applicability of the methods appropriate to simple random sampling that are the subjects of the following chapters.

When the population we wish to make inferences about can be readily divided into different subpopulations, it is often advantageous to select sub-samples separately from each subpopulation. It may be the case that some subpopulations are more intensively sampled than others. If so, an element in a subpopulation that is more intensively sampled has a greater chance of inclusion in the overall sample than an element in a subpopulation that is less intensively sampled. This violates the fundamental principle of simple random sampling. When we select separate random subsamples from different subpopulations, we call it *stratified random sampling*. An instance in which we might apply stratified random sampling would be in selecting samples of ceramic sherds for raw material sourcing from each of the eight known households at an excavated site. Each household would be a *sampling stratum*, and we would make inferences about where the ceramics in each household were made based on independent samples. All eight samples might later be combined for purposes of making inferences about where ceramics at the settlement as a whole were made. Stratified random sampling is discussed in Chapter 16.

When the elements of the population we wish to make inferences about are not available individually for selection, we often use sampling strategies based on spatially defined selection units. If the population we are interested in, for example, consists of the lithic artifacts in a particular site, we can likely select a simple random sample of artifacts only if the site has already been excavated and the artifacts are in a laboratory or museum where we can select sample members individually. If the site has not been excavated we may nonetheless wish to obtain a random sample of lithic artifacts from it.

This could be done by excavating small test pits in a number of different locations to recover some of the artifacts that lie buried in the site's deposits. If the locations of those test pits were randomly selected (for example, by establishing a grid system over the site area and randomly selecting which grid units to excavate), then the resulting artifacts would still be a random sample of artifacts from the site. They would not, however, be a simple random sample, because the elements in the sample (that is, lithic artifacts) were not individually selected. It was grid units that were individually selected randomly, and so we do have a simple random sample of grid units. But it is a population of lithic artifacts, not a population of grid units, that we wish to make inferences about in the present instance. Lithic artifacts were selected, not individually, but rather in small groups or *clusters*, each cluster being those lithic artifacts contained in the deposits in one excavated grid unit. We then have not a simple random sample of artifacts but rather a *cluster random sample* of artifacts. Cluster samples, like stratified samples, are within the reach of statistical tools—the ones taken up in Chapter 17.

Several other terms have come to be used here and there in the archaeological literature for nonrandom ways of selecting samples—"haphazard sampling," "grab sampling," "judgmental sampling," "purposive sampling," and the like. These are not well-established terms that have clear meanings with precise statistical implications. They refer to the explicit or implicit application of a variety of nonrandom selection criteria. In some circumstances it may be justifiable to treat such samples as if they were random samples, but such treatment must be applied with caution, and specific justification for it is always required.

For example, a surface collection made at an archaeological site is sometimes described as a haphazard sample or a grab sample, probably meaning that field workers walked around an area and picked up haphazardly an assortment of the things they saw. The resulting sample is then sometimes used as a basis for making inferences about the population from which this haphazard sample was selected, presumably the entire set of artifacts on the surface of the site at the time. This approach is likely to produce a sample that systematically misrepresents the population in certain respects. For example, a haphazard surface collection of this sort is likely to contain a higher proportion of large artifacts than the population did, simply because they will be more noticeable. As a consequence, if the sample is used to make inferences about the average size of artifacts on the surface of the site, the inferences will be inaccurate. Similarly, if the sample is used to make inferences about the proportions of different artifact classes on the surface of the site, classes of artifacts that tend to be large will seem to be more abundant than they really were.

Much the same could be said about color and other characteristics that affect the visibility of artifacts lying on the ground. Even more subtly, a haphazard surface collection like this may have higher proportions of unusual things and correspondingly lower proportions of common things than does the artifact population from which it comes, as a consequence of a subconscious tendency to collect things that strike the eye as especially different from most of the other things being seen.

This haphazard sample is *biased* because the elements in it were selected in a way that makes the sample systematically different from the population in certain respects. There is no statistical technique for eliminating such bias once the sample has been selected. The appropriate statistical tool for avoiding sample bias is random selection of the sample, and this tool must be used at the time the sample is selected. It cannot be applied retroactively. Haphazard or grab samples are simply not the same as random samples.

Judgmental or purposive samples are also likely to be biased. These terms tend to refer to samples selected by looking over the range of elements in a population and specifically deciding to include certain elements in the sample and exclude others. Obviously, whatever the criteria involved in the selection, the resulting sample will be biased with respect to those characteristics. Suppose, for example, that an archaeologist wishes to study the residential remains at a site where the locations of individual households are marked on the surface by small mounds. He or she might decide to thoroughly excavate those mounds that show the highest densities of artifacts on their surfaces on the theory that their excavation will produce greater numbers of artifacts. The result, of course, will be a sample of house mounds with a substantially higher number of artifacts than average for the site as a whole. Such a sample could clearly not be used to make inferences about the average density of artifacts in house mounds at the site.

Still more insidious are other possibly related factors. The higher artifact densities that caused mounds to be included in the sample might be the result of, say, the greater wealth of certain households and consequently more intensive disposal of used and broken objects near them. Thus the sample might systematically misrepresent the wealth of households at the site and therefore be biased in this regard as well. Inferences about the entire population in regard to such characteristics as the proportions of different artifact or ecofact classes related to wealth would be systematically erroneous.

Once again, the moral of the story is that random selection of the elements in a sample is the only way to ensure that a sample is unbiased. Random samples are the only ones that we can be sure are unbiased, because their method of selection specifically avoids conscious and subconscious biases of the kinds just discussed. Since bias refers to the systematic application of criteria that result in an unrepresentative sample, we know that biased samples are unrepresentative in certain ways. The reverse is not, however, true. That is, the absence of bias from random samples does not guarantee that each and every one will accurately represent its parent population. We can never be entirely certain that a sample accurately represents the population from which it was selected (unless we study the entire population). Any given random sample may be very representative or very unrepresentative. Biased samples are known to be unrepresentative in some ways. We can (and will in the next few chapters) assess the probability that a random sample is unrepresentative, something we cannot do with any except random samples.

USE OF NONRANDOM SAMPLES

Most of the statistical tools discussed in this book require us to assume that we are working with random samples. Most archaeological data, however, were (and continue to be) produced with nonrandom sampling procedures that result in biased samples. This, on the surface, would suggest that statistical tools are not applicable to most archaeological data. And this, indeed, is the conclusion at which some archaeologists have arrived. The situation, however, is simultaneously more and less serious than this.

First the bad news. The difficulty of making inferences about populations from samples guaranteed to be biased in certain ways is not unique to statistical means of making inferences. We cannot reliably infer anything about the average size of artifacts on the surface of a site from a collection that overrepresents large artifacts. This is true whether our means of inference are statistical or purely intuitive. We are simply unable to make conclusions by any means about the average size of artifacts in the population on the basis of such a sample. It is no solution to avoid statistical approaches and rely strictly on subjective impressions or any other kind of inference, because all kinds of inferences are affected in precisely the same way by sampling bias. Thus, the need for unbiased samples is by no means a need imposed by the arcane rules of statistical inference. It is fundamental to inference of any kind, and we ignore it only at our own peril whether we use statistical tools or not.

Now the good news. This is not another cautionary tale ending at the nihilistic conclusion that reliable interpretation of archaeological remains is impossible. Too many such tales have already appeared in the archaeological literature. Thorough understanding of the nature of sample bias and careful application of common sense can make inferences about populations from samples possible. Moreover, clear thinking about this issue, stimulated by efforts to apply statistical techniques, can be carried over productively into the arena of nonstatistical inference making, leaving us in position to make more reliable conclusions in other ways as well.

The effects of known or possible bias in sample selection can (and must) be evaluated with particular reference to at least two specific ways in which biased samples might still be used. First, a sample that is biased in one respect is not necessarily useless for any and all purposes since it may not be biased in other respects. If a case can be made that the bias in sample selection is unrelated to some other characteristic of the population, then the sample might be appropriate for making inferences about that other characteristic. Second, two samples selected with the same bias may still be usefully compared even with regard to the characteristic involved in the selection bias. Here the case that must be made is that the bias operated similarly enough in the selection of both samples to have had a very similar impact on both. The two samples then might be unrepresentative of their parent populations in precisely the same way, making some kinds of conclusions from comparing them reliable. Judgments in such instances are likely to involve ad hoc reasoning more than the application of general rules or principles, and the process is, perhaps, made clearest through examples rather than abstract discussion.

Example: A Haphazard Surface Collection

In the instance of a haphazard collection of artifacts from the surface of a site, very small artifacts are almost certainly not collected as frequently as larger ones are, simply because they are considerably less noticeable. (If we want to be truly honest about it, the same could probably be said of most artifact samples recovered from screens during excavation.)

We may well not be interested in inferring anything about the average size of artifacts, however. In many such instances it is, on the other hand, of considerable interest to infer the relative proportions of different ceramic types in the parent population. If we can make the case that sherd size is unrelated to ceramic type, then even a sample selected with bias in regard to sherd size can be used to make such inferences reliably. Only if some ceramic types tended systematically to break into substantially smaller pieces than others and therefore systematically to be underrepresented in the sample would sampling bias on account of size affect inferences about the proportions of different ceramic types. The possibility of a relationship between sherd size and ceramic type could be evaluated empirically before proceeding to use a haphazard artifact collection as the basis for such an inference. Similarly, other possible kinds of bias resulting from haphazard sample selection can be enumerated and their impacts on particular kinds of inferences that we are interested in assessed.

Even if we determine that sample bias makes our inference about proportions of different ceramic types suspect in this instance, this suspect inference might still be usefully compared with similar inferences concerning other sites based on samples selected with the same biases. As long as the operation and strength of the bias in sample selection can be supposed to be the same for all the samples, then the inaccurate inferences may be, in effect, comparably inaccurate. A sample that underrepresents a particular type can be usefully compared to another sample that underrepresents the same type to the same extent. Such comparisons are quite often the ultimate objective in working with type proportions anyway. It may have no absolute significance to us that a particular type represents, say, 30% of the ceramics on the surface at some site—only that this 30% is greater than the figure of 15% obtained from another site. For purposes of such conclusions it makes no difference at all, finally, that the truly accurate numbers might be 36% and 18%, respectively, instead of 30% and 15%. For comparative purposes, then, sampling bias may, in effect, cancel itself out when it affects all samples in the same way and to the same extent.

Example: A Purposive Obsidian Sample

Many archaeologists have been faced with a sampling decision in regard to raw material sourcing. Obsidian artifacts, for example, from many parts of the world can be linked to sources of raw material through chemical fingerprinting. The necessary analyses, however, are so expensive that it is usually possible to identify only a portion of the obsidian recovered from a site. In this situation some archaeologists have looked over all the obsidian obtained from the site and selected as many pieces as they can afford to analyze, intentionally including artifacts of as many different colors and appearances as possible. The justification for this procedure has usually been that it provides the greatest chance of including material from the largest possible number of different sources since material from different sources may differ visually as well as chemically. Since some sources may be represented by only a few pieces in a large population, there is a very good chance that those sources might not turn up at all in a random sample of modest size—hence the interest in including in the sample for analysis pieces of very unusual appearance.

The sampling procedure is thus biased, systematically overrepresenting in the sample artifacts of unusual appearance. If there is, indeed, some relation between appearance and source location, this bias makes the sample irretrievably inappropriate for inferring what proportion of the artifacts were made with materials from the different sources. To see why, one can imagine drawing a sample of 4 marbles from a jar with 97 black marbles, 1 blue marble, 1 red marble, and 1 green marble. Clearly the best representation of the full range of different colors in the population can be achieved by purposely selecting a sample consisting of one marble of each color. That sample, however, could not then be used to estimate the proportions of different-colored marbles in the population. Observing that the sample consisted of 25% black, 25% blue, 25% red, and 25% green would lead directly to the inference that the proportions in the population were also 25% of each color, but we know the real proportions to be 97% black, 1% blue, 1% red, and 1% green. The proportions in the sample were determined not really by any characteristic of the population but rather entirely by the biased sampling procedure.

A sample of obsidian artifacts selected in such a way for source analysis simply cannot be used to make any inference at all about the proportions of material acquired from different sources, no matter how useful it may be for obtaining as long a list as possible of different sources exploited. The sample may well be used for other inferences insofar as it is possible to argue that the bias in sample selection does not relate to these other inferences. For example, such a sample might be used to study whether material from different sources tended to be worked in different ways. The selection bias would, superficially at least, not appear to relate to this issue.

The only way to be absolutely certain that sampling bias has no effect on inferences made, of course, is to be certain that sample selection is entirely free from bias. Random sampling is the appropriate technique for avoiding bias in sample selection, and should be applied whenever unbiased samples are needed (even when statistical means of making inferences are not contemplated). To the extent that the case can be made, however, that bias in the selection of already existing samples does not affect the specific inferences being made, then those inferences can be considered reliable. And that means quite literally *to the extent that the case can be made*—to that extent; no more and no less. Like so many other things in life, sampling bias is not a matter of black or white but of varying shades of gray. Clear, careful thinking may convince us that the risk of sample bias, insofar as a particular inference is concerned, is minimal, even though the sample at hand is egregiously biased in other regards. If we can postulate a number of specific ways that bias might affect the inferences we are interested in and then empirically rule out all these possibilities, then the case for disregarding bias (and treating an existing sample as if it were a random sample for a particular purpose) may be quite convincing. If, in contrast, we simply ignore the possibility of such problems, then any inferences made must be viewed with suspicion.

This is not a perspective on sampling bias that is often expressed in the archaeological literature or elsewhere, and it certainly runs counter to the rules laid down in many statistics textbooks—particularly those of the cookbook persuasion. Statistics books that emphasize memorizing rules (the "Ours not to reason why" approach) are likely to forbid the application of most of the techniques discussed in this book to any sample not strictly randomly selected (by which they mean with a table of random numbers or similar procedure to preclude bias). This would mean that these techniques could not be applied to the vast majority of archaeological data now in existence. Worse yet, since, as we have seen above, sampling bias affects not just statistical inferences but any kind of inferences from samples to populations, we would not be in a position to make any inferences at all from these data.

Presumably, those who adopt such a stringent position will not be much attracted to archaeology and will not be reading this book. The rest of us will find it necessary to do the best we can. (It has been pointed out by others that archaeology falls not among the hard sciences but instead among the difficult sciences.) First and foremost, we can take advantage of the proper techniques to guard against sampling bias to the maximum extent possible. When we cannot do this (as, for example, when we wish to learn what can be learned from previously selected samples), we must use our wits to assess how serious an impact sampling bias may have on particular inferences and structure our analyses so as to minimize it. Sometimes we will make inferences that must be taken with caution because some impact from sampling bias cannot be ruled out. If these inferences prove interesting, they may justify further data collection to see if they hold up even with unbiased samples.

THE TARGET POPULATION

The previous discussion may imply that adoption of strict random sampling procedures could resolve the issue of sampling bias in archaeology once and for all by avoiding it entirely. An even stronger view, sometimes argued in the archaeological literature, is that sampling bias is best avoided by abandoning sampling altogether in favor of studying entire populations. Neither of these solutions, however, will work in archaeology because the *target population* about which we wish to make inferences is seldom fully available either to study in its entirety or to select a sample from.

At a regional scale, at least some sites in any region are likely to be unavailable for study because they have been covered by modern urban concentrations, obscured by recent sedimentation, carried away by erosion, or otherwise destroyed or made inaccessible. Thus even a regional survey that is complete in the sense of systematically covering the entire surface of the region does not have access to the complete population of sites that we need to study. A sample selected by the strictest random sampling procedures remains not a sample of all the archaeological sites ever left but a sample of those sites that remain accessible for selection. Similarly, at a smaller scale, the vast majority of archaeological sites are not intact and completely preserved but are only partial, with some sectors destroyed or inaccessible to study. Thus, whether we study entire archaeological populations or take random samples of them, the populations truly available for study or sampling do not precisely correspond to the populations about which we wish to make inferences.

Random sampling puts us in position to make inferences about the population the sample was drawn from, and, of course, study of an available population provides us with conclusions about that available population. If that available population, however, was only the part of an important site that had not been washed away by the adjacent river, we are faced with the difficult question of how to attempt to characterize the entire meaningful site. There is no simple and straightforward solution to this difficulty, just as there is no simple and straightforward solution to the problem of making inferences from biased samples. The most common response by archaeologists to this difficulty is simply to ignore it. This response is clearly conceptually inadequate, although a number of famous archaeologists have built successful careers on it. Another common response is to pretend that the missing part of the site contained what we hoped to find but didn't find in the part that remains. This is just plain unconvincing.

Fundamentally, the difficulty of not being able to study or sample from the population we are truly interested in parallels the problem of sampling bias. The population available to be studied or sampled is, in effect, itself a sample from the target population—one selected by quite possibly very biased procedures (whatever processes destroyed or made inaccessible the portion we cannot now study). It is because archaeologists are so frequently in this position that they are forced to sample in one way or another. Often the entire population available for study is already a sample. We thus cannot escape the complexities of sampling and the issue of sampling bias, no matter how we try.

Whether we select samples ourselves, work with data from samples other people selected, or study entire available populations, we still must wrestle the sampling bias problem to ground as best we can if we propose to do archaeology at all. This means using our understandings of sampling and sampling bias to say as much about the representativeness of a sample as possible, using statistical tools presented in this book and/or using nonstatistical and probably ad hoc reasoning applicable to specific instances.

Even when we can apply random sampling procedures to an available population that corresponds well to the target population about which we wish to make inferences, avoiding sampling bias does not guarantee a representative sample, as discussed above. It only gives us the best chance of a representative sample and enables us to assess the probabilities of its unrepresentativeness.

In any of these cases, then, some of the inferences we make about entire populations from the samples we can study will be correct and some will be incorrect. Some will be incorrect because the population from which we could select a sample did not represent very accurately the population about which we wish to know. Some will be incorrect because the sample we study does not accurately represent the population from which it was selected. Although related, these are two different sources of error. The first must be dealt with on an ad hoc basis with cleverness and common sense. Random sampling and the statistical tools discussed in the next few chapters can help us with the second by telling us roughly what percentage of our inferences are incorrect for this reason. We cannot, however, determine specifically which ones are incorrect. Without these tools we can say even less. If we are careful and diligent, most of our conclusions will be correct, but it is unrealistic to hope to make correct inferences 100% of the time, no matter how careful we are to eliminate sampling bias (and other inaccuracies). Finally, confidence in our ultimate conclusions is best reinforced by finding consistent patterns in the majority of multiple independent inferences. When such consistent patterns are recognized, it should make us willing to set aside inconsistent inferences as possible consequences of sampling error (of either of the two kinds mentioned above).

To those who are concerned that I have taken here too cavalier an attitude toward the importance of random sampling in statistical (or other) inference, I can only say that I see no other way to proceed in most of the situations that practicing archaeologists must actually face. The course advocated here is to try to rule out all the likely ways in which a sample may be biased. If it seems likely that a sample may be unbiased, then it is worth setting our quite proper worries about sampling bias aside for the moment at least and going ahead to see what inferences about the population in which we are interested our sample may lead to. If it does lead to interesting inferences about the population, then our worries about sampling bias must return as the proverbial grains of salt with which our conclusions must be taken. If we are fairly confident that the sample we are working with can be taken to be unbiased, then we can be fairly confident about the conclusions concerning a population that we make on the basis of that sample. If we think the sample we are working with might be biased, then whatever conclusion we arrive at about a population on the basis of that sample must be taken with a correspondingly large grain of salt.

Practitioners of most other disciplines do not find these issues as troublesome as archaeologists do, because they are usually interested in studying target populations that are much more accessible for study than those of the archaeologist. They can often afford to ignore results from samples that may be biased, and simply go back to the field or laboratory for a more carefully selected sample. Much of the sampling bias in archaeological samples, however, is not so easily avoided. We must learn, then, to avoid sample bias whenever we can (as by selecting truly random samples) and to live and work productively alongside it when we must. When the discrepancies between our real target population and the population actually available to be sampled are truly large, excessive finickiness about sample selection procedures begins to be like straightening deck chairs on the Titanic. We need to be careful and thoughtful in deciding when straightening deck chairs is a worthwhile activity and when our attention would better be directed to the lifeboats.

Much of this discussion anticipates the statistical techniques discussed in Chapters 8–10 and may not make too much sense to those who do not already have some inkling of them. The issues raised will come up again repeatedly in this book, however, and the discussion in this chapter lays out the reasons for approaching them in the way that we will. We will return to them in the last chapter as well.

PRACTICE

1. Imagine that you have made an intensive surface collection at the Keeney Knob site. The following Saturday night you happen to meet an amateur archaeologist who used to own a farm at Stony Point. Later on, he lets you study the large collection of lithic artifacts he made on his farm before they built the shopping center and obliterated all trace of the archaeological site. You immediately recognize that the lithics from Stony Point are precisely contemporaneous with the ones you have collected at Keeney Knob, and you are eager to compare the artifacts from the two sites. First, you

would like to know whether the Keeney Knob and Stony Point lithic assemblages have similar or different proportions of projectile points. Of the artifacts in your surface collection from Keeney Knob, 14% are projectile points; of the collection from Stony Point, 82% are projectile points. Second, you are interested in the raw materials from which projectile points were made at the two sites. Of the Keeney Knob projectile points, 23% are obsidian and 77% are chert; of the Stony Point projectile points, 6% are obsidian and 94% are chert. You recognize, however, that you have a potential problem of sampling bias in making use of these comparisons. How would you assess this problem, and what would you do about it? Can you make any use at all of these comparisons? Can you be more confident about conclusions from one of them than from the other? Why?

2. You have data from haphazard surface collections at a series of neolithic sites in the Velika Morava River valley. They were made by an archaeological team during a field season in 1964 before the area was flooded by a reservoir. If your hypothesis about the beginnings of grain cultivation in the region is correct, the sites in river bank locations should have substantially larger proportions of stone hoes than the sites set back from the river. What worries about sampling bias would you have to face in using the data from the 1964 Velika Morava survey to investigate your hypothesis? How would you face these worries? How much confidence would you place in conclusions you arrived at about the proportions of stone hoes at different sites in this region, based on the 1964 survey? Why?

Chapter 8

Different Samples from the Same Population

The discussion in Chapter 7 dealt with the fact that sometimes random samples represent the populations from which they are drawn very accurately and sometimes they don't. Random selection is no guarantee of representativeness. Random sample selection does, however, make it possible to apply some very powerful tools for assessing how likely it is that a sample is unrepresentative to a particular degree. This is because, with random samples, we can say something about how often particular degrees of unrepresentativeness occur on average.

ALL POSSIBLE SAMPLES OF A GIVEN SIZE

In order to understand this we must consider the many possible different random samples that can be drawn from a single population. Table 8.1 contains the measurements (in cm) of the diameters of 17 post holes from excavations at a single site. The measurements have been arranged in ascending order to make them easier to examine. We will consider these 17 measurements a population of measurements from which we wish to draw a sample. This is, of course, an exceedingly small-scale example. Samples themselves are likely to consist of far more than 17 measurements, and the populations from which the samples are drawn are even larger. But this small

Table 8.1. Diameter Measurements for a Small Population
of Post Holes[a]

Post hole number	Diameter (cm)	Post hole number	Diameter (cm)
1	10.4	10	13.2
2	10.7	11	13.7
3	11.1	12	14.0
4	11.5	13	14.3
5	11.6	14	15.0
6	11.7	15	16.4
7	12.2	16	18.4
8	12.6	17	20.3
9	12.9		

[a] $N = 17$; $\mu = 13.53$ cm; $\sigma = 2.73$ cm.

example enables us to see in operation principles that it would be almost impossible to observe in an example of large enough scale to be more realistic.

This population, then, consists of 17 post holes, the diameters of which have been measured. We will use the capital letter N to indicate the number of elements in a population; thus, for this example $N = 17$. The mean post hole diameter for the 17 post holes in the population is 13.53 cm, and we will use μ (the Greek lower-case letter mu) to stand for the mean of the population. Thus $\mu = 13.53$ cm. The standard deviation of a population is represented by σ (the Greek lower-case letter sigma), so in this example $\sigma = 2.73$ cm.

We will begin by considering the smallest possible sample, a sample of 1. The lower-case letter n represents the number of elements in a sample, just as N represents the number of elements in a population. We will consider all the possible samples of 1 ($n = 1$) that could be drawn from this population of 17 post holes ($N = 17$). It is easily seen that there are 17 possible different samples of 1 that might be selected. We might randomly select post hole No. 1, or No. 2, or No. 3, . . . or No. 17. Whichever sample of 1 we happened to select, we could calculate the mean post hole diameter in that sample and use it to infer the mean post hole diameter for the entire population. Our best guess for the population mean is always the sample mean. In order to distinguish these two means in equations we use μ to stand for the population mean, as in Table 8.1, and $\overline{X}$ to stand for the sample mean. Thus, the best estimate of μ is always $\overline{X}$.

If our sample consisted of post hole No. 1, we would guess that the mean post hole diameter in the population was 10.4 cm, since 10.4 cm is the mean of a sample consisting of the single observation 10.4 cm. If our sample consisted of post hole No. 2, we would guess that the population mean was 10.7 cm, and so on. From the 17 different samples of 1 that we might select we could make 17 different guesses at the population mean. Some of these guesses would be very close (as for the samples consisting of post hole No. 10 or post hole No. 11). Other guesses would be much farther off (as for the samples consisting of post hole No. 1 or post hole No. 17). This example shows clearly that some samples represent the population from which they are drawn relatively accurately, and others do not.

The largest possible error in estimating the population mean occurs when the sample of 1 consists of post hole No. 17. On the basis of this sample we would guess that the population mean was 20.3 cm, an error of 6.77 cm. This is certainly a regrettably large error. Moreover, such a maximum error will occur fairly often in drawing samples of 1. Fully 1/17 (5.9%) of the total number of different samples of 1 that could be drawn from this population would consist of post hole No. 17. Thus, if we were to select samples of 1 repeatedly from this population, 5.9% of these samples would consist of post hole No. 17, causing us to make such an erroneous guess at the population mean. Sampling in this way, then, we would make an error as large as 6.77 cm 5.9% of the time.

If we needed to estimate the mean post hole diameter in this population with an error no greater than 3.0 cm, we could figure out how often we would succeed and how often we would fail in this example. Of the 17 possible samples of 1 that we might select, 3 samples would result in estimates of the population mean with an error greater than 3.0 cm, and 14 samples would result in estimates with an error of 3.0 cm or less. (The samples consisting of post hole No. 1, post hole No. 16, and post hole No. 17 would have means different from the known population mean by more than 3.0 cm.) Thus, 82.4% of the samples of 1 would provide us with estimates as accurate as we needed, but 17.6% of them would not.

If we selected samples of 1 over and over again, then 82.4% of the time we would get a sample yielding an acceptably accurate estimate of the population mean, and 17.6% of the time we would get a sample yielding an unacceptably inaccurate estimate. (At least this is the case if each of these different samples is equally likely to occur, which of course is true if the samples are randomly selected.) These percentages translate directly into probabilities for any *single* instance of drawing a sample of one. That is, if 82.4% of the samples of 1 that we might draw would yield an acceptably accurate estimate of the population mean, then the *probability* of arriving at an acceptably accurate estimate in any single instance of drawing a sample of 1 would be 82.4% (or .824).

Stating the probability of occurrence of a single event in this manner means nothing more than stating the percentage of occurrence of that single event in a long sequence of repeated trials. We are accustomed to making such statements as, for example, when we say that the probability that a tossed coin will turn up heads is 50%. In saying this, we mean that when we toss a coin repeatedly, 50% of the time the result is heads. On a single toss the result will be either heads or tails, not half heads and half tails, but the probability of heads on a single toss is 50% because in repeated trials, 50% of the time the result will be heads and 50% of the time the result will be tails. This way of talking about probabilities is largely a matter of common sense and well established in common speech, but its importance to statistics is such that it merits explicit statement here.

In the example of drawing samples of 1 from a population of 17 post holes, then, we would achieve successful (that is, acceptably accurate) results 82.4% of the time. We would fail to attain the accuracy needed 17.6% of the time. If this success rate is not high enough, common sense tells us that we might do better with a larger sample.

ALL POSSIBLE SAMPLES OF A LARGER GIVEN SIZE

Suppose we selected samples of 2 post holes each from the population of 17 post holes. The range of possible results here is much larger. Our sample of 2 might consist of post hole No. 1 and post hole No. 1. (We are sampling here with replacement as discussed in Chapter 7.) Or our sample might consist of post hole No. 1 and post hole No. 2; or of post hole No. 1 and post hole No. 3; or of post hole No. 2 and post hole No. 3; and so on. In all there are 153 possible different samples of 2 post holes that could be selected from the population of 17 post holes (with replacement). If the samples were randomly selected, each of these 153 possible different samples of 2 would be equally likely to occur on any given drawing.

Of these 153 possible different samples of 2, some, of course, would give us estimates of the population mean with the level of accuracy we need (an error no greater than 3.0 cm) and some would not. It is not too difficult to determine which ones. Clearly, the sample consisting of post hole No. 1 and post hole No. 1 would give a mean diameter of 10.4 cm, more than 3.0 cm in error. The next smallest possible sample mean would come from the sample consisting of post hole No. 1 and post hole No. 2. Here the mean diameter for the sample would be 10.55 cm. This is 2.98 cm less than the true population mean, so the error is acceptably small. There is, then, only one possible sample of 2 that would give an estimate of the population mean more than 3.0 cm too small.

At the other end of the scale, sample means more than 3.0 cm higher than the population mean would be produced by all of the following samples of 2:

post holes No. 17 and No. 17 ($\bar{X} = 20.30$ cm)
post holes No. 17 and No. 16 ($\bar{X} = 19.35$ cm)

post holes No. 17 and No. 15 ($\bar{X}$ = 18.35 cm)
post holes No. 17 and No. 14 ($\bar{X}$ = 17.65 cm)
post holes No. 17 and No. 13 ($\bar{X}$ = 17.30 cm)
post holes No. 16 and No. 16 ($\bar{X}$ = 18.40 cm)
post holes No. 16 and No. 15 ($\bar{X}$ = 17.40 cm)
post holes No. 16 and No. 14 ($\bar{X}$ = 16.70 cm)

All the remaining possible samples of 2 that we might select would yield means no more than 3.0 cm different from the true population mean and would thus be acceptably accurate.

In sum, then, of the 153 possible different samples of 2 that we might select from the population of 17 post holes, 1 sample would yield an unacceptably low estimate of the population mean, 8 samples would yield unacceptably high estimates of the population mean, and 144 samples would yield acceptably accurate estimates of the population mean. Thus 144/153 (94.1%) of the time we would achieve successful (that is, acceptably accurate) results, and 5.9% of the time we would fail to achieve acceptable accuracy in estimating the population mean. Our probability of success on any given sample selection, then, is substantially greater with samples of 2 (acceptable accuracy 94.1% of the time) than it is with samples of 1 (acceptable accuracy 82.4% of the time). Samples of 2 that give unacceptably inaccurate results are more unusual than are samples of 1 that give unacceptably inaccurate results. Thus it is less likely that any particular random sample of 2 that we might select would give us unacceptably inaccurate results than was the case for samples of 1. The probability that any particular random sample of 2 yields unacceptably inaccurate results is 5.9% (or .059) in contrast to the probability of 17.6% (or .176) that any particular random sample of 1 would yield unacceptably inaccurate results. This is because such unrepresentative samples are more unusual among all possible samples of 2 than among all possible samples of 1.

If we extend the example to samples of 3, the same trend continues. There are 2,601 possible different samples of 3 that we might select from the population of 17 post holes. Of these, the following would yield estimates of the population mean more than 3.0 cm too low:

post holes No. 1, No. 1, and No. 1 ($\bar{X}$ = 10.40 cm)
post holes No. 1, No. 1, and No. 2 ($\bar{X}$ = 10.50 cm)

In addition, the following samples of 3 would yield estimates of the population mean more than 3.0 cm too high:

post holes No. 17, No. 17, and No. 17 ($\bar{X}$ = 20.30 cm)
post holes No. 17, No. 17, and No. 16 ($\bar{X}$ = 19.67 cm)

post holes No. 17, No. 17, and No. 15 ($\bar{X} = 19.00$ cm)
post holes No. 17, No. 17, and No. 14 ($\bar{X} = 18.53$ cm)
post holes No. 17, No. 17, and No. 13 ($\bar{X} = 18.30$ cm)
post holes No. 17, No. 17, and No. 12 ($\bar{X} = 18.20$ cm)
post holes No. 17, No. 17, and No. 11 ($\bar{X} = 18.10$ cm)
post holes No. 17, No. 17, and No. 10 ($\bar{X} = 17.93$ cm)
post holes No. 17, No. 17, and No. 9 ($\bar{X} = 17.83$ cm)
post holes No. 17, No. 17, and No. 8 ($\bar{X} = 17.73$ cm)
post holes No. 17, No. 17, and No. 7 ($\bar{X} = 17.60$ cm)
post holes No. 17, No. 17, and No. 6 ($\bar{X} = 17.43$ cm)
post holes No. 17, No. 17, and No. 5 ($\bar{X} = 17.40$ cm)
post holes No. 17, No. 17, and No. 4 ($\bar{X} = 17.37$ cm)
post holes No. 17, No. 17, and No. 3 ($\bar{X} = 17.23$ cm)
post holes No. 17, No. 17, and No. 2 ($\bar{X} = 17.10$ cm)
post holes No. 17, No. 17, and No. 1 ($\bar{X} = 17.00$ cm)
post holes No. 17, No. 16, and No. 16 ($\bar{X} = 19.03$ cm)
post holes No. 17, No. 16, and No. 15 ($\bar{X} = 18.37$ cm)
post holes No. 17, No. 16, and No. 14 ($\bar{X} = 17.90$ cm)
post holes No. 17, No. 16, and No. 13 ($\bar{X} = 17.67$ cm)
post holes No. 17, No. 16, and No. 12 ($\bar{X} = 17.57$ cm)
post holes No. 17, No. 16, and No. 11 ($\bar{X} = 17.47$ cm)
post holes No. 17, No. 16, and No. 10 ($\bar{X} = 17.30$ cm)
post holes No. 17, No. 16, and No. 9 ($\bar{X} = 17.20$ cm)
post holes No. 17, No. 16, and No. 8 ($\bar{X} = 17.10$ cm)
post holes No. 17, No. 16, and No. 7 ($\bar{X} = 16.97$ cm)
post holes No. 17, No. 16, and No. 6 ($\bar{X} = 16.80$ cm)
post holes No. 17, No. 16, and No. 5 ($\bar{X} = 16.77$ cm)
post holes No. 17, No. 16, and No. 4 ($\bar{X} = 16.73$ cm)
post holes No. 17, No. 16, and No. 3 ($\bar{X} = 16.60$ cm)
post holes No. 17, No. 15, and No. 15 ($\bar{X} = 17.70$ cm)
post holes No. 17, No. 15, and No. 14 ($\bar{X} = 17.23$ cm)
post holes No. 17, No. 15, and No. 13 ($\bar{X} = 17.00$ cm)
post holes No. 17, No. 15, and No. 12 ($\bar{X} = 16.90$ cm)
post holes No. 17, No. 15, and No. 11 ($\bar{X} = 16.80$ cm)
post holes No. 17, No. 15, and No. 10 ($\bar{X} = 16.63$ cm)
post holes No. 17, No. 14, and No. 14 ($\bar{X} = 16.76$ cm)
post holes No. 16, No. 16, and No. 16 ($\bar{X} = 18.40$ cm)
post holes No. 16, No. 16, and No. 15 ($\bar{X} = 17.73$ cm)
post holes No. 16, No. 16, and No. 14 ($\bar{X} = 17.27$ cm)
post holes No. 16, No. 16, and No. 13 ($\bar{X} = 17.03$ cm)
post holes No. 16, No. 16, and No. 12 ($\bar{X} = 16.93$ cm)
post holes No. 16, No. 16, and No. 11 ($\bar{X} = 16.83$ cm)
post holes No. 16, No. 16, and No. 10 ($\bar{X} = 16.67$ cm)

post holes No. 16, No. 16, and No. 9 ($\bar{X} = 16.57$ cm)
post holes No. 16, No. 15, and No. 15 ($\bar{X} = 17.07$ cm)
post holes No. 16, No. 15, and No. 14 ($\bar{X} = 16.60$ cm)
post holes No. 16, No. 15, and No. 14 ($\bar{X} = 16.60$ cm)

Thus 2 of the 2,601 possible samples of 3 would yield unacceptably low estimates and 49 would yield unacceptably high estimates. The acceptable accuracy rate would be 2550/2601, or 98.0%. The probability of selecting a random sample of 3 from this population of post holes that would yield an unacceptably inaccurate estimate of the population mean, then, is only 2.0% (or .02). This is because random samples of 3 with sample means very different from the mean of the population from which they were selected are fairly unusual (representing only 2.0% of the possible samples). It is thus very likely (not certain but very likely) that any particular sample of 3 that we might select from the population would represent the population with the accuracy we decided was needed in this example.

We could continue this example by considering the 44,217 possible different samples of 4 that could be selected, but the point should by now be clear. The larger the random sample is, the greater the chance that it represents the population from which it is selected with acceptable accuracy. Other things being equal, it is the size of the sample that governs its likely representativeness. Larger samples are more often representative of their parent populations than small samples. But, as has been emphasized above, this condition provides *no guarantee* of representativeness. The most unrepresentative sample of 3 in this example consists of post hole No. 17 selected three times. This sample is just exactly as unrepresentative as the most unrepresentative sample of 1 (consisting of post hole No. 17). But such unrepresentative samples occur far less frequently among larger samples than among smaller samples.

The number of errors of more than 3.0 cm in estimating the mean in the population of 17 post hole diameters also depends on the spread of the population. If there are many post holes much larger or much smaller than the mean, then the number of samples producing unacceptably inaccurate results increases. If this does not initially make sense to you, go back to the example population given in Table 8.1 and change post holes 1, 2, and 3 to 9.0 cm, 9.4 cm, and 9.8 cm, respectively. Start counting up how many samples of 1, 2, and 3 there would be with means more than 3.0 cm different from 13.53 cm. The bigger the spread in the population, the more samples there will be whose means are not acceptably close to the true population mean (for any given definition of "acceptably close").

The chance of making badly erroneous inferences about populations on the basis of samples, then, is less with larger samples, although a small risk of serious error remains even with large samples. The chance of making badly erroneous inferences about populations on the basis of samples is also less when the population is homogeneous (a batch with a small spread) and greater when the population is highly variable (a batch with a larger spread). In this specific example, in which we know exactly what the population is like, and we established (even if arbitrarily) what "acceptable" accuracy was, we could easily figure the percentages of samples that would yield acceptable and unacceptable results. What we need now is a means of generalizing the observations that we can make in this specific example.

THE "SPECIAL BATCH"

The key to general application of the specific observations we made in the example above lies in a very special batch of numbers. *This "special batch" consists of the means of all the possible different samples of a given size that could be drawn from a given population.* Let's consider this in terms of the previous example.

For a sample size of 1 (that is, for $n = 1$), there are 17 different random samples that could be selected from our example population of 17 post holes. Each of the 17 samples has its own sample mean ($\bar{X}$). The special batch would consist of these 17 sample means. We found above that 17.6% of these 17 sample means were more than 3.0 cm different from the real population mean, and they were therefore classified as unacceptably unrepresentative samples. Unacceptably unrepresentative samples of 1 were thus a bit unusual, making up only 17.6% of the special batch, but we would not call them extremely rare. The clear majority of the samples of 1 that we could select from this population would represent it with sufficient accuracy for our present purposes, but an uncomfortably large proportion of the samples of 1 we might select would be unacceptably inaccurate.

For $n = 2$, there are 153 different random samples that could be selected from our example population of 17 post holes. Each of these 153 samples has its own sample mean ($\bar{X}$). The special batch would consist of these 153 sample means. Samples so unrepresentative that their means differed by more than 3.0 cm from the population mean were more unusual in terms of this special batch, making up only 11.8% of the possible samples of 2 that could be selected from this population.

For $n = 3$, there are 2,601 different random samples that could be selected from our example population of 17 post holes. Each of these 2,601 samples has its own sample mean ($\bar{X}$). The special batch would consist of these 2,601 sample means. Unacceptably unrepresentative samples were even more unusual among samples of 3, making up only 2.0% of the special batch.

And we could go on. For a given population and for any given sample size, there is a special batch consisting of the means of all the different samples of that size that could be selected randomly from that population. This special batch, then, consists of all the possible results we could obtain in estimating the given population's mean on the basis of a sample of the given size. And this special batch is the key to determining just how unusual it would be to draw an unacceptably unrepresentative sample of a certain size from the given population. The unusualness of an unacceptably unrepresentative sample (in terms of the special batch) enables us to specify the probability that any specific sample of a given size that we might randomly select from a given population will be unrepresentative.

THE STANDARD ERROR

We have just been using the notion of unusualness in a way very similar to the way we used it in Chapter 4—unusualness of a number in terms of the batch of numbers to which it belongs. Since the numbers we have been discussing are the means of samples of particular sizes, the comparison batch has been the batch consisting of the means of samples of a given size from a given population, that is, the special batch. In Chapter 4 we talked about more general procedures for evaluating the unusualness of a number in terms of its batch, procedures based on numerical indexes of the level and spread of the batch. We could apply just such procedures to this effort to discuss unusualness of sample results in terms of the special batch. In order to do so we would need to know the level and spread of the special batch. We could, of course, find out the level and spread of the special batch by selecting all possible samples of a given size and working directly with the batch, but this is obviously preposterous. It would be considerably more work than just studying whatever we wanted to study in the whole population, and so sampling would offer no advantage. It turns out that there are much easier ways to find out about the special batch.

It can be shown mathematically that *the mean of the special batch is the same as the mean of the population from which the samples were drawn*. This, of course, is quite apparent in the case of samples of 1. The special batch for samples of 1 is exactly the same batch of numbers as the population, since each sample is the same as one number in the population. The mean of the population, then, has to be the same as the mean of the special batch. It turns out that this is true even when $n > 1$ (that is, even when the sample size is greater than 1).

If we can say that the mean of this special batch is the same as the mean of the population from which the samples are drawn, then we can say that the mean of the means of all the possible samples of a given size that can be drawn from a given population is the same as the mean of that population. These two statements are synonymous because the special batch *is* the means of all the possible samples of a given size that can be drawn from a given population.

You can actually think this through fairly easily for yourself if you want to, without need of formal mathematical proofs. If we select all possible samples of any given size, each number in the population occurs an equal number of times in all the samples taken together (however many times that may be—it depends on the sample size). The mean of all the sample means is also the mean of all the numbers in all the samples, taken as one immense undivided batch. Since all numbers in the population occur the same number of times in all the samples taken together, this immense batch is simply the original population reduplicated many times over, and its mean will be the same as the mean of the original population. Each number has simply been added in many times, but then the total has been divided by a much larger number, reflecting precisely how many times each number has been added in.

It can also be shown mathematically that *the standard deviation of the special batch is the standard deviation of the given population divided by the square root of the given number of elements in the sample.* The truth of this is, once again, obvious when the given sample size is 1. The standard deviation of the special batch is the standard deviation of the population divided by the square root of 1 (the given sample size). Since the square root of 1 is 1, the standard deviation of the special batch is the same as the standard deviation of the population when the given sample size is 1. This is not surprising, since the special batch is the same as the population when the given sample size is 1. This same relationship, however, between the standard deviation of the special batch and the sample size and the standard deviation in the population holds true for any given sample size.

The standard deviation of the special batch is such an important number that it has its own special name. It is the standard error. *The standard error, then, is the standard deviation of the batch consisting of the means of all the different samples of a given size that could be selected from a given population.* The equation for standard error is

$$SE = \frac{\sigma}{\sqrt{n}}$$

where

SE = standard error;
σ = standard deviation of the population;
n = number of elements in the sample.

We are now in position to specify a numerical index of level and a numerical index of spread for the special batch so as to discuss the unusualness of particular samples in a general and efficient way. The numerical indexes we have specified, however, are two very ill-behaved ones. Neither mean nor standard deviation is at all resistant to the effect of outliers or skewness. Here we are in luck, however, because for samples of relatively large size it can also be shown mathematically that the shape of the special batch is normal. Since normal shapes are single peaked and symmetrical, we know that the mean and standard deviation will be useful numerical indexes of level and spread, and we do not have to worry about the fact that they are not resistant. Relatively large sample size, in this instance, can be taken to mean more than about 30. This characteristic of the special batch (having a normal shape for relatively large sample size) is also of pivotal importance. It is called the *central limit theorem*.

To summarize, in this section we have conceived of a special batch of numbers that consists of the means of all the different samples of a given size that could be drawn from a given population. This special batch is known in more formal statistical terminology as the *sampling distribution of the mean*, but we will continue to refer to it here simply as the special batch. Three properties of the special batch have been noted. First, the mean of the special batch is the same as the mean of the population from which the samples are selected. Second, the standard deviation of the special batch, known as the standard error, is $\sigma/\sqrt{n}$. And third, the shape of the special batch is normal as long as the sample size is over about 30.

These three properties of the special batch give us rather complete information about its characteristics. Without having to actually select and manipulate all possible samples of a given size, we can determine the level (mean), spread (standard deviation), and shape (single peaked, symmetrical, normal) of the special batch. In the next chapter we will put the special batch and its characteristics to general use in assessing the unusualness of particular samples.

Chapter 9

Confidence and Population Means

The major difficulty in putting the properties of the special batch discussed in Chapter 8 to use is that we had to know a good deal about the population from which the sample was drawn in order to specify the characteristics of the special batch. We knew that the mean of the special batch was the same as the mean of the population and that the standard deviation of the special batch (that is, the standard error) was the standard deviation of the population divided by the square root of the number in the sample. In real life, however, we do not know either the mean or the standard deviation of the population from which our sample is drawn. Indeed, those are precisely the things we are likely to be trying to infer on the basis of a sample. Thus we must find a way to use the special batch without first knowing these characteristics of the entire population.

In this chapter we will extend the notion of unusualness of a sample to apply to the more realistic situation in which instead of having one population and all the possible samples from it we have one sample and consider the possible populations it might have come from. We will start by asking the question, "How unusual would it be for the sample we actually have to come from a population with a particular mean?" And we will proceed to ask that question about a number of different possible parent populations for our sample.

GETTING STARTED WITH A RANDOM SAMPLE

Let's suppose that we have a random sample of 100 projectile points drawn from a much larger population of projectile points, whose mean length we wish to know. This random sample of 100 projectile points has a mean length of 3.35 cm and a standard deviation of 0.50 cm. Such a situation may occur in real life when, for example, we have surveyed a region intensively and made systematic surface collections at all the sites encountered. To keep the logic simpler, let's suppose that study of these collections revealed occupation of the region during only a single prehistoric period. We decide to take all the projectile points recovered in these collections (100 points altogether) as a random sample from the population consisting of all the projectile points made by the prehistoric inhabitants of the region during the single period during which the region was occupied.

Our sample is not technically a random sample, but we might decide to treat it as one, at least for estimating the mean projectile point length in the population. In order to make this decision we would need to consider the collecting procedures used in the field as well as the processes by which projectile points are brought to the surfaces of sites and become available for collection. These latter processes include the full range of things that happen to projectile points from the time they are discarded to the time they are found. If, in considering all these processes, we can find no reason to believe that projectile points of different lengths will be affected in substantially different ways (or at least that whatever such effects may be, they apply equally to this sample and to other samples with which we wish to compare this sample), then we would proceed to treat this sample as a random sample with respect to projectile point length. The legitimacy of any conclusions we make about projectile point length in the population, of course, is dependent on this decision. We must recognize the possibility in using these conclusions that, at some time in the future, they might be invalidated if we were to discover that the sample had been biased with respect to projectile point length in some way we had not thought of.

This procedure may seem risky, but, as discussed in Chapter 7, the only alternative is simply not to make conclusions about projectile point length in the larger population. Whatever statements we make about, say, Late Woodland projectile points in general are based on precisely such logic, whether those statements are statistical in nature or purely subjective impressions.

Archaeologists have always made such general statements about large and vaguely defined populations on the basis of samples not randomly selected. And such statements, even when statistics have been in no way involved, are based on treating the sample at hand as if it were not biased even when we cannot show conclusively that bias is absent. This approach is no more risky when it serves as the foundation for statistical statements than when it serves as the foundation for subjective impressions. Indeed it is less risky. This is because the statistical techniques we are about to apply only assume that the sample is unbiased; they do not assume that it accurately represents the population from which it came, only that it is not systematically biased. Subjective generalizations assume not only that the sample upon which they are based is unbiased but also that the sample provides completely accurate representation—a stronger assumption, and one much more difficult to justify.

Archaeologists are not the only scientists in this situation. We are all comfortable using such figures as the mean heights of adult males and adult females in the United States. We seldom even think about where such figures come from. Clearly they do not involve measuring the heights of all adult males and all adult females in the country. The figures are based on a much smaller sample. Even that is not technically a random sample of all adult males and all adult females in the country. It was a sample from a much smaller subpopulation *taken to accurately represent the larger population*. No one ever actually assigned numbers to every adult male and every adult female in the United States, randomly selected a sample, and set out to measure every individual in the sample. Much smaller and more accessible populations were taken to accurately represent the nation's population at large after careful consideration and elimination of the ways in which such populations might be biased samples.

In exactly the same way, archaeologists do not need to be able to number sequentially and randomly select a sample from all the projectile points made in a particular period in a particular region in order to characterize this large and vaguely defined population. Archaeologists can (and must) argue that the projectile points lying on the surface at a given moment are an unbiased subgroup of that larger population (with respect to certain characteristics at least) and that the 100 projectile points recovered on survey are an unbiased sample from that subgroup. This is the way sampling of such large and vaguely defined populations is customarily done in many disciplines. The conclusions produced are reliable only to the extent that the assumption that the sample is unbiased can be justified. If this is in doubt, then this doubt remains as a doubt about the validity of the conclusion reached.

As long as we have digressed from the topic at hand to such a lengthy discussion of the real-life implications of the assumptions of sampling, we might as well specify one terminological point as well. Large and vaguely defined populations like the one we are dealing with here are referred to in statistics as *infinite populations*. This does not mean that they are truly infinite, just that they are very large and not precisely defined. (We will see as we continue to discuss the notion of infinite populations that infinity is a much smaller thing to a statistician than to an astronomer.)

FROM WHAT POPULATIONS MIGHT THE SAMPLE HAVE COME?

Once we've satisfied ourselves that we are willing to treat the sample we have as if it were a random sample (at least for purposes of argument), we can begin to consider from what kind of population the sample might have come. Recall that our sample of 100 projectile points had a mean length of 3.35 cm and a standard deviation of 0.50 cm. For large populations and large samples, the sample mean is the same as the population mean more often than it is any other one figure. Similarly, the sample standard deviation is the same as the population standard deviation more often than it is any other one figure. Thus our best estimate is that the population of projectile points from which this sample was selected has a mean length of 3.35 cm and a standard deviation of 0.50 cm.

We know, however, that samples do not always have exactly the same mean as their parent populations, so we wonder just how much confidence we should have in this estimate. Put another way, just how likely is it that this estimate is incorrect? Put more fully, just how likely is it that this estimate is incorrect by enough to matter? The addition of that last phrase is an important practical matter of precision. We almost certainly do not need to worry about the possibility that the real population mean might be 3.350000001 cm as opposed to 3.350000000 cm. This difference of 0.000000001 cm is clearly not enough to matter. It is almost certainly well beyond the capability of our measuring instruments even to detect such a difference. But the point is that we do not seek infinite precision even if it were possible—it wouldn't matter. Being incorrect by enough to matter is what we have to worry about. Probably 0.01 cm or even 0.1 cm is not enough to worry about. Maybe even 0.4 cm or 0.5 cm is not a large enough error in estimating the population mean to worry us seriously.

The question of necessary precision is not one of applying statistical rules of precision. Rather it is a substantive question involved with why we want to know what the mean length of projectile points in this population is. For statistical purposes, then, we take whatever decision is made about necessary precision as a given because that decision is based on substantive concerns outside the realm of statistics. For example, our reason for wanting to know the mean length of projectile points in our region may be to compare this length with the mean length for another region in an effort to determine something about differences in hunting practices. In this case, a difference of 0.1 cm would likely not be taken as meaningful in that it would seem too small to be reflecting a meaningful difference in hunting practices. A difference of 0.5 cm might, on the other hand, be meaningful, if a substantive case could be made for what, specifically, it would indicate.

Turning back to the sample that we have, we have already guessed that it most likely came from a population with a mean length of 3.35 cm (the same as the sample mean). But we know that there is no guarantee that it came from such a population. Our sample might have come from a population with a mean length greater or less than 3.35 cm, possibly even from a population with a mean length much greater or much less than 3.35 cm. We can begin to think about how likely this is by considering various specific populations from which our sample might have come. For each specific population we imagine that our sample might have come from, we will need to think of the special batch consisting of the means of all possible samples of 100 from that population.

For starters, let's imagine our sample might have come from a population with a mean length of 3.25 cm. How unusual would it be to get a sample like ours (that is, with a mean of 3.35 cm and a standard deviation of 0.50 cm) from a population with a mean of 3.25 cm? What would the special batch consisting of the means of all possible samples of 100 from a population with a mean of 3.25 cm look like? We know that the mean of this special batch would be the same as the population mean, that is, 3.25 cm. We know that the shape of this special batch would be approximately normal because of the central limit theorem and because 100 is a fairly large sample. We only lack knowledge of the spread of the special batch, but we know that the spread of the special batch is given by the equation

$$SE = \frac{\sigma}{\sqrt{n}}$$

Since we have no better recourse, we will continue to use the standard deviation of the sample (0.50 cm) as our best estimate of the standard deviation in the parent population. Thus

$$SE = \frac{0.50 \text{ cm}}{\sqrt{100}} = \frac{0.50 \text{ cm}}{10} = 0.05 \text{ cm}$$

Figure 9.1 illustrates the special batch consisting of the means of all the possible samples of 100 that could be drawn from a population with a mean

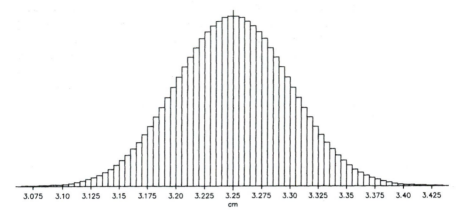

Figure 9.1. The special batch consisting of the means of all the possible samples of 100 that could be selected from a population with a mean of 3.25 cm and a standard deviation of 0.50 cm.

of 3.25 cm and a standard deviation of 0.50 cm. This is simply a histogram, like those discussed in Chapter 1. Clearly, samples with means close to 3.25 cm are much more common than are samples with means far from 3.25 cm. Figure 9.2 illustrates this same special batch in a more common and useful manner. Instead of a histogram with specific intervals represented by vertical bars, the heights of the bars are represented by a smooth curve joining the center points of the tops of the bars. This allows us to use the horizontal

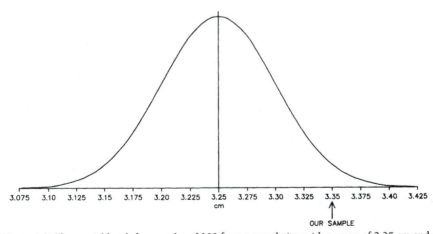

Figure 9.2. The special batch for samples of 100 from a population with a mean of 3.25 cm and a standard deviation of 0.50 cm.

scale as the truly continuous measurement scale that it is instead of breaking it up into awkward intervals. The height of the curve above any point along the horizontal scale, then, represents the frequency with which samples with a particular mean occur, in just the same way that the height of a bar on the corresponding histogram represents the frequency of occurrence of samples with a mean falling in a particular interval. Representing the shape of a batch with a normal distribution in this way is so common in statistics that the entire concept is often referred to in a kind of shorthand as the "normal curve."

For a given mean (in this case 3.25 cm) and a given standard deviation (in this case the standard error, which is the standard deviation of the special batch, or 0.05 cm) there is one and only one specific normal distribution, and Figure 9.2 is it. Figure 9.2 is thus a picture of the special batch consisting of the means of all possible samples of 100 that can be selected from a population with a mean of 3.25 cm and a standard deviation of 0.50 cm. We can use this picture to place our sample, with a mean of 3.35 cm, in context with all other possible samples. The position of our sample in this distribution is indicated in Figure 9.2. At the point corresponding to our sample, the normal curve is fairly low, indicating that samples with a mean of 3.35 cm do occur among the possible samples of 100 from a population with a mean of 3.25 cm, but they do not occur very frequently—not nearly as frequently, for example, as samples with means closer to 3.25 cm. Our sample is fairly unusual, then, in the context of all the possible samples from a population with a mean of 3.25. It is therefore possible, but not very likely, that our sample came from a population with a mean of 3.25 cm.

We can do the same thing for other populations from which our sample might possibly have come. For instance, how likely is it that our sample came from a population with a mean length of 3.20 cm? Figure 9.3 illustrates the special batch consisting of the means of all possible samples of 100 that could be selected from a population with a mean of 3.20 cm and a standard deviation of 0.50 cm. The level of the normal curve at the point corresponding to our sample in Figure 9.3 is extremely low. Thus our sample, with its mean of 3.35 cm, would be extremely unusual among samples of 100 selected from a population with a mean of 3.20 cm. It is therefore very unlikely (although not entirely impossible) that our sample came from such a population.

How likely is it that our sample came from a population with a mean of 3.30 cm? Figure 9.4 illustrates the special batch consisting of the means of all possible samples of 100 that could be selected from a population with a mean of 3.30 cm and a standard deviation of 0.50 cm. The level of the normal curve at the point corresponding to our sample in Figure 9.4 is fairly high. Thus there are a good many samples like ours among those possible to select from a population with a mean of 3.30 cm. Therefore it is relatively likely that our sample could have come from such a population.

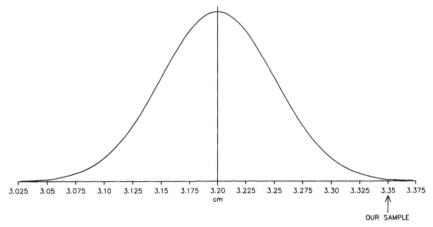

Figure 9.3. The special batch for samples of 100 from a population with a mean of 3.20 cm and a standard deviation of 0.50 cm.

Finally, Figure 9.5 illustrates the special batch corresponding to the population with a mean of 3.35 cm—the population that is a more likely parent population for our sample than any other single population. We could imagine continuing to try out many more possible parent populations in this way and constructing a new curve from the results of these trials. This new curve would indicate how likely it was that each of the possible parent

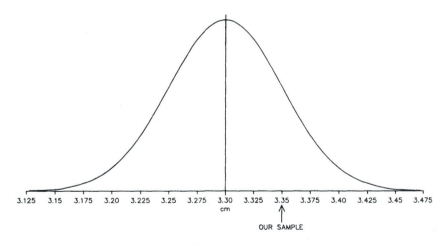

Figure 9.4. The special batch for samples of 100 from a population with a mean of 3.30 cm and a standard deviation of 0.50 cm.

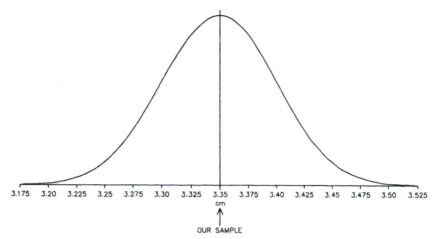

Figure 9.5. The special batch for samples of 100 from a population with a mean of 3.35 cm and a standard deviation of 0.50 cm.

populations was indeed the population from which our sample was drawn. It turns out that if we carried out this procedure the curve we would construct would have exactly the same parameters as the curve illustrated in Figure 9.5. In effect what we have done is turn the curve in Figure 9.5 inside out to produce the curve in Figure 9.6.

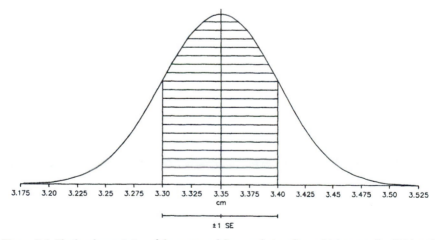

Figure 9.6. The batch consisting of the means of the populations from which a sample of 100 with a mean of 3.35 cm and a standard deviation of 0.50 cm might have come. The majority of the means lie within 1 standard error of the sample mean, but a substantial number of means are larger or smaller than this.

Figure 9.5, again, represents the special batch composed of the means of all the possible samples of 100 that could be selected from a population with a mean of 3.35 cm and a standard deviation of 0.50 cm. It thus represents the unusualness of the various samples that could be selected from this population and therefore the probability of selecting any one of them from this population. Figure 9.6, on the other hand, represents the means of the possible populations that a sample of 100 with a mean of 3.35 cm and a standard deviation of 0.50 cm might have been drawn from and therefore the probability that this sample was selected from any particular one of them. This batch, represented in Figure 9.6, has exactly the same level, spread, and shape as the special batch that we have been discussing. That is, just like the familiar special batch, this second batch has a mean that is the same as the sample mean; it has a standard deviation that is $\sigma/\sqrt{n}$, or the standard error; and its shape is normal.

CONFIDENCE VERSUS PRECISION

We can look at Figure 9.6 and quickly say that a good many of the populations that our sample might have come from have means between 3.30 cm and 3.40 cm. (These are the populations that fall within 1 standard error of the mean of our sample.) According to the shape of the special batch, however, a good many of the possible populations have means outside that range. Thus we are only moderately confident that the population our sample came from has a mean between 3.30 cm and 3.40 cm. We say this because populations with means less than 3.30 cm or greater than 3.40 cm are relatively numerous among the possible populations. It would not strain credulity at all to imagine selecting a sample with a mean of 3.35 cm and a standard deviation of 0.50 cm from a population with a mean less than 3.30 cm or greater than 3.40 cm. Figure 9.6 shows us that such a thing would happen with some frequency. Thus our sample probably came from a population with a mean between 3.30 cm and 3.40 cm, but there is a very real chance that it might not have. It means the same thing to say, "The probability is moderate that our sample came from a population with a mean of 3.35 cm ± 0.05 cm."

Suppose we are not satisfied with the lack of confidence we have in the statement that the population our sample came from probably has a mean between 3.30 cm and 3.40 cm. We can speak more confidently, but only by reducing the level of precision of our statement. We could say that the population our sample came from has a mean between 3.25 cm and 3.45 cm, and be somewhat more confident that our statement is true. This statement is illustrated by Figure 9.7, where the clear majority of the possible populations have means that fall between 3.25 cm and 3.45 cm. It seems quite likely that

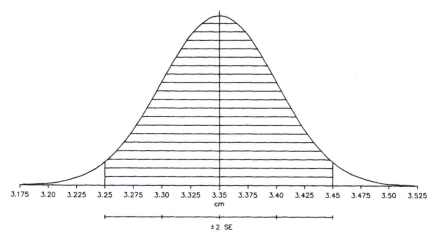

3.175   3.20   3.225   3.25   3.275   3.30   3.325   3.35   3.375   3.40   3.425   3.45   3.475   3.50   3.525

cm

± 2 SE

Figure 9.7. The batch consisting of the means of the populations from which a sample of 100 with a mean of 3.35 cm and a standard deviation of 0.50 cm might have come. The vast majority of the means lie within 2 standard errors of the sample mean.

our sample comes from a population with a mean somewhere in this range. Relatively few of the possible populations fall outside the range. Thus it would be fairly unusual to select a sample like ours (with a mean of 3.35 cm and a standard deviation of 0.50 cm) from a population with a mean less than 3.25 cm or greater than 3.45 cm. The probability that our sample came from a population with a mean less than 3.25 cm or greater than 3.45 cm is low. Correspondingly, the probability that our sample came from a population with a mean between 3.25 cm and 3.45 cm is high. Thus we might say something like, "There is a high probability that our sample came from a population with a mean of 3.35 cm ± 0.10 cm." This statement indicates greater confidence than the statement at the end of the previous paragraph, but it is a less precise statement.

The twin notions of confidence and precision are familiar to us in common colloquial speech, although we usually don't think of them directly. If I intend to make quite sure I will arrive for an appointment at a precise time, I might say, "I will be there at 4 o'clock." Customs of punctuality vary, but I am not likely to say that unless I feel quite confident that I will arrive within about 5 minutes of 4 o'clock. If my arrival depends on how heavy traffic is en route, I am more likely to say, "I will be there about 4 o'clock," a less precise statement, indicating that I might be 10 or 15 minutes early or late. If I envision still more imponderable interference with my schedule, I might say, "I will be there sometime around 4 o'clock," indicating still less precision, say between 3:30 and 4:30.

I could communicate similar messages by varying the confidence implied in my statements. Instead of saying "I will be there about 4 o'clock," I could say, "I will probably be there at 4 o'clock." The former statement encourages the listener to think of a period of 20 minutes or so during which my arrival can be expected. The latter statement instead encourages the listener to imagine the precise moment of 4 o'clock, but not to have too much confidence that I will be present then. Both statements convey approximately the same message, but I might use them in different contexts. If I am going to a meeting with a colleague, which will begin when I arrive, I would say "I will be there about 4 o'clock," thinking of the range of time during which the meeting can be expected to begin. If, on the other hand, I am going to a lecture scheduled to start at 4 o'clock whether I am there or not, I would say, "I will probably be there at 4 o'clock," imagining how likely it is that I will be present at the precise time the lecture can be expected to begin. It is usually a trade-off between speaking with precision and speaking with confidence. Other things being equal, the more precision we speak with, the lower our confidence; and the more confidence we speak with, the less precise our statements. Only in unusual circumstances am I able to say, "I *will* be there at 4 o'clock sharp," emphasizing that I am speaking with both high confidence ("I *will*") and high precision ("4 o'clock sharp").

The statistical statements we are making about the kind of population our sample came from work in exactly the same way. We can either indicate very high confidence that the population has a mean in a somewhat imprecise range of values or indicate a population mean with greater precision but

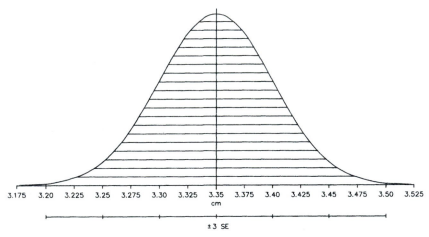

Figure 9.8. The batch consisting of the means of the populations from which a sample of 100 with a mean of 3.35 cm and a standard deviation of 0.50 cm might have come. Only a few means are more than 3 standard errors from the sample mean.

lower confidence that we're correct. Figure 9.8 continues the progression begun in Figures 9.6 and 9.7. It illustrates a still less precise statement, but one that can be made with great confidence. Almost all the possible populations that a sample like ours (of 100 elements with a mean of 3.35 cm and a standard deviation of 0.50 cm) could come from have means that fall in the range between 3.20 cm and 3.50 cm. Very few of the possible populations have means less than 3.20 cm or greater than 3.50 cm. It would be quite unusual to select a sample of 100 with a mean of 3.35 cm and a standard deviation of 0.50 cm from a population with a mean less than 3.20 cm or greater than 3.50 cm. Thus it is very unlikely that our sample came from a population with a mean less than 3.20 cm or greater than 3.50 cm. It is very likely that our sample came from a population with a mean between 3.20 cm and 3.50 cm. We could say, "The probability is very high that the population our sample came from has a mean of 3.35 cm ± .15 cm."

PUTTING A FINER POINT ON PROBABILITIES—STUDENT'S *t*

The notions of approximate probabilities we have been using thus far can be extended to much more precise and useful ways of assessing probabilities on the basis of how unusual a particular result would be in the context of all the possible results. We have used the approximate height of the normal curve (and thus the shaded areas enclosed by it in Figures 9.6, 9.7, and 9.8) to judge roughly how unusual (and thus how improbable) it would be for our sample to have been selected from populations with means falling in different ranges. These ranges of possible means are called *error ranges*, or *confidence intervals*. They are most often expressed as a "±" quantity following the mean. Figure 9.6 illustrates an error range of ± 1 standard error, Figure 9.7 illustrates an error range of ± 2 standard errors, and Figure 9.8 illustrates an error range of ± 3 standard errors. We concluded above that we are very confident that the mean of the population our sample came from lies within the ± 3 standard error range (Figure 9.8); we are fairly confident that the mean of the population our sample came from lies within the ± 2 standard error range (Figure 9.7); and we have only modest confidence that the mean of the population our sample came from lies within the ± 1 standard error range (Figure 9.6).

The exact *levels of confidence* we have in these three statements of differing precision can be found by calculating the exact areas "under the normal curve" in Figures 9.6, 9.7, and 9.8. Student's *t* distribution provides us with these exact areas. The key to use of Student's *t* is in numerical indexes of level and spread (in this case the mean and standard deviation) used to measure the unusualness of a particular number in a batch. The relevant batch is the special batch, whose mean is the same as the mean of our sample and whose standard deviation is the standard error. (Be sure not to confuse the standard deviation of the sample or of the population with the standard deviation of the special batch. The standard deviation of the special batch is the standard error of the sample.) Student's *t*, then, provides a detailed description of the shape of the special batch for us to use.

Figure 9.7, for example, illustrates an error range (3.35 cm ± 0.10 cm) consisting of 2 standard errors. We already recognized that it is very likely that the population our sample came from has a mean that falls within this error range. Table 9.1 allows us to say just what we mean by "very likely" in the following manner. First we must determine the row of the table to use, based on the size of the sample. The left-hand column indicates the *degrees of freedom*, which are equivalent to one less than the number of elements in the sample $(n - 1)$. For the moment, we will just take this notion of degrees of freedom (often abbreviated *d.f.*) on faith. For our sample, $n - 1 = 99$. There is no row corresponding exactly to 99 degrees of freedom, so we will use the row for 120 *d.f.* We are looking for the exact level of confidence associated with an error range of 2 standard errors, so we read across that row looking for 2. In the fourth column we find 1.98 (which we'll take as close enough to 2 for the moment).

The fourth column is headed 95% confidence. This means that 95% of the possible populations (represented by the shaded area "under the normal curve" in Figure 9.7) that our sample could come from lie within 1.98 standard errors of the mean of our sample. Thus, when we say that it is "very likely" that our sample came from a population with a mean of 3.35 cm ± 0.10 cm, what we mean more precisely is that there is about a 95% probability that our sample came from such a population. We are 95% confident that our sample came from a population with a mean of 3.35 cm ± 0.10 cm. We are *not certain* that our sample came from a population with a mean of 3.35 cm ± 0.10 cm, but the probability that this is the case is 95%.

Since the probability that our sample came from a population with a mean between 3.25 cm and 3.45 cm is 95%, the probability that it came from a population with a mean less than 3.25 cm or greater than 3.45 cm is 5%. (This has to be true since the probability that it came from one or the other of these groups is 100%.) Since a normal shape is symmetrical, this 5% is evenly distributed in both "tails" of the distribution. There is a 2.5% probability that our sample came from a population with a mean less than 3.25 cm and a 2.5% probability that our sample came from a population with a mean greater than 3.45 cm. When we provide an error range of about 2 standard errors, then, as we have done here, we are speaking at a *95% confidence level*. This follows directly from the observation that a number that falls 2 standard deviations or more away from the mean in its batch is a very unusual number in terms of its batch. Specifically, only about 5% of the numbers in a normally distributed batch fall this far from the mean.

Table 9.1. Student's t Distribution[a]

CONFIDENCE (%)

(NO. OF STANDARD ERRORS)

SIGNIFICANCE (%)

(NO. OF STANDARD ERRORS)

Confidence	50%	80%	90%	95%	98%	99%	99.5%	99.8%	99.9%
	.5	.8	.9	.95	.98	.99	.995	.998	.999
Significance	50%	20%	10%	5%	2%	1%	0.5%	0.2%	0.1%
	.5	.2	.1	.05	.02	.01	.005	.002	.001

Degrees of freedom

1	1.000	3.078	6.314	12.706	31.821	63.637	127.32	318.31	636.62
2	.816	1.886	2.920	4.303	6.965	9.925	14.089	22.326	31.598
3	.765	1.638	2.353	3.182	4.541	5.841	7.453	10.213	12.924
4	.741	1.533	2.132	2.776	3.747	4.604	5.598	7.173	8.610
5	.727	1.476	2.015	2.571	3.365	4.032	4.773	5.893	6.869
6	.718	1.440	1.943	2.447	3.143	3.707	4.317	5.208	5.959
7	.711	1.415	1.895	2.365	2.998	3.499	4.020	4.785	5.408
8	.706	1.397	1.860	2.306	2.896	3.355	3.833	4.501	5.041
9	.703	1.383	1.833	2.262	2.821	3.250	3.690	4.297	4.781
10	.700	1.372	1.812	2.228	2.764	3.169	3.581	4.144	4.537
11	.697	1.363	1.796	2.201	2.718	3.106	3.497	4.025	4.437
12	.695	1.356	1.782	2.179	2.681	3.055	3.428	3.930	4.318
13	.694	1.350	1.771	2.160	2.650	3.012	3.372	3.852	4.221
14	.692	1.345	1.761	2.145	2.624	2.977	3.326	3.787	4.140
15	.691	1.341	1.753	2.131	2.602	2.947	3.286	3.733	4.073
16	.690	1.337	1.746	2.120	2.583	2.921	3.252	3.686	4.015
17	.689	1.333	1.740	2.110	2.567	2.898	3.222	3.646	3.965
18	.688	1.330	1.734	2.101	2.552	2.878	3.197	3.610	3.922
19	.688	1.328	1.729	2.093	2.539	2.861	3.174	3.579	3.883
20	.687	1.325	1.725	2.086	2.528	2.845	3.153	3.552	3.850
21	.686	1.323	1.721	2.080	2.518	2.831	3.135	3.527	3.819
22	.686	1.321	1.717	2.074	2.508	2.819	3.119	3.505	3.792
23	.685	1.319	1.714	2.069	2.500	2.807	3.104	3.485	3.767
24	.685	1.318	1.711	2.064	2.492	2.797	3.091	3.467	3.745
25	.684	1.316	1.708	2.060	2.485	2.787	3.078	3.450	3.725
30	.683	1.310	1.697	2.042	2.457	2.750	3.030	3.385	3.646
40	.681	1.303	1.684	2.021	2.423	2.704	2.971	3.307	3.551
60	.679	1.296	1.671	2.000	2.390	2.660	2.915	3.232	3.460
120	.677	1.289	1.658	1.980	2.358	2.617	2.860	3.160	3.373
∞	.674	1.282	1.645	1.960	2.326	2.576	2.807	3.090	3.291

[a]Adapted from Table 3 in *Introduction to Contemporary Statistical Methods* by Lambert H. Koopmans (Boston: Duxbury Press, 1987).

   Every error range (or confidence interval) expressed in terms of stand-ard errors corresponds to a specific confidence level. (The terms *confidence interval* and *confidence level* are too close for comfort, considering that they refer to two rather different concepts. Thus the term *error range* is used here in preference to *confidence interval*.) An error range of ± 3 standard errors, as illustrated in Figure 9.8, corresponds to approximately 99.8% confidence. Reading across the row in Table 9.1 that corresponds to 120 *d.f.*, as we did before, looking for 3 brings us to the next-to-last column, where 3.160 is relatively close to 3. This column is headed 99.8% confidence. Thus when we concluded, on the basis of Figure 9.8, that it is very likely that our sample comes from a population with a mean of 3.35 cm ± 0.15 cm, that "very likely" actually meant a probability of around 99.8%. There is only about a 0.2% probability that the population our sample came from has a mean less than 3.20 cm or greater than 3.50 cm. Once again, since a normal shape is symmetrical, that means about a 0.1% probability that the population our sample came from has a mean less than 3.20 cm and about a 0.1% probability that it has a mean greater than 3.50 cm.

   Finding the confidence level associated with a 1 standard error range is a little more difficult with Table 9.1. Reading across the row for 120 *d.f.*, we see values that skip from 0.677 to 1.289. The confidence level corresponding to a 1 standard error range thus falls between these two columns. The columns are headed 50% confidence and 80% confidence. For a large sample such as this, the confidence level corresponding to a 1 standard error range actually is about 66%.

ERROR RANGES FOR SPECIFIC CONFIDENCE LEVELS

   In some circumstances when we express inferences about population means as error ranges, we simply use 1 standard error as the error range. This has become the normal practice with radiocarbon dates, to choose an example with which archaeologists are comfortable—even those most uneasy about statistics. The error ranges given with radiocarbon dates are under-stood by convention to be 1 standard error (and we are accustomed to calling them "error ranges" rather than the more statistically traditional "confidence intervals"). These ranges are arrived at by application of precisely the princi-ples we have just discussed to the sample of emitted particles counted in the laboratory. We can thus apply exactly the same kinds of statements we have just been making to radiocarbon dates. We are moderately confident that the date of death of the carbon atom population from which came the sample that decayed while in the laboratory counter falls within the 1 standard error range specified. More accurately, the probability that the real date of the carbon falls within that range is 66%. This still leaves quite a substantial risk that the real date falls outside that range. If we double the usual range (to arrive at 2 standard errors), we are stating the date less precisely (with an error range twice as large), but we can be 95% confident that the real date falls within that larger range. These ranges, of course, as we have all been warned since we first read introductory textbooks, refer only to the risk of error resulting from the process of measuring the quantity of carbon 14, and are in addition to whatever loss of confidence results from risks like mistaken context, contamination, and the like.

It is worth noting that the standard practice widely accepted by archaeologists in radiocarbon dating is an example of precisely the argument made in Chapter 7 and earlier in this chapter for using samples not strictly randomly selected as a basis for inferences about populations we are interested in. Radiocarbon date error ranges are based on a random sample of carbon 14 atoms (those that decay while the specimen is in the counter). But this sample is drawn from a population of carbon 14 atoms rolled up in an aluminum foil packet in the field through no rigidly random sampling procedure. And the inference made about that population of atoms on the basis of a random sample from it is readily extended to characterize something much broader than that aluminum foil packet. If we are responsible about it, we always bear in mind the risks of mistaken context, contamination, and so on that might invalidate the extension of that inference to the phenomenon we are really interested in, but we do not let the mere existence of such risks paralyze our use of a very powerful dating technique. We can follow just such procedures with many other kinds of samples as well—recognizing the possibility that they may be biased samples from the populations we are really interested in, but that it is worth going ahead and studying them anyway because the possibility of such bias may never be absolutely eliminated.

The 1 standard error range, in any event, has considerable precedent behind it in archaeology because it is the standard for radiocarbon dating. Sometimes an error range of 2 standard errors is used when an author is willing to speak less precisely in exchange for higher levels of confidence. Providing error ranges like this has one principal disadvantage. The corresponding confidence levels are not entirely self-evident. We found above that, in the case of our example sample of 100 projectile points, a 1 standard error range corresponds to about 66% confidence. In the same case a 2 standard error range provides 95% confidence, and a 3 standard error range provides about 99.8% confidence.

These confidence levels can be used as rules of thumb, but they do not hold true if the sample under consideration is small. Suppose our sample had consisted of only six projectile points. We would have needed to use the row in Table 9.1 for 5 d.f. $(n - 1)$. In this row, we find a t value of approximately 2

Be Careful How You Say It

When you estimate the mean of a population on the basis of a sample and provide an error range for the estimate, it is essential to specify the confidence level as well. Virtually the only exception to this rule is for radiocarbon dates where the convention of providing error ranges of ± 1 standard error is firmly established. The conclusion reached in the example discussed at length in the text might, for example, be stated, "We estimate, on the basis of our sample, that the projectile points used by the inhabitants of our region during the one prehistoric period when the region was occupied had a mean length of 3.35 cm ± .08 cm (at the 90% confidence level)." Alternatively, we might say, "Our sample indicates 90% confidence that the mean length of projectile points in our region was 3.35 cm ± .08 cm." It is not incorrect to say, "Our sample indicates 90% confidence that the mean length of projectile points in our region was between 3.27 cm and 3.43 cm." It is probably better, however, to express the error range as a ± figure associated with the mean. Stating only the maximum and minimum values of the range encourages some people to think that all values within that range are equally likely estimates, and that values outside the range are not possible. We know, however, that the mean itself is the single most likely estimate, and that there is some possibility that the "correct" population value actually lies outside whatever error range is expressed.

---

in the column corresponding to 90% confidence rather than the 95% confidence we found before.

To provide error ranges at a fixed level of confidence irrespective of sample size, it is necessary to use the *t* table to determine exactly how many standard errors are required for the desired confidence level. In the case of the sample of 100 projectile points with a mean length of 3.35 cm and a standard deviation of 0.50 cm, we might want to express our estimate of the mean projectile point length in the population with an error range at the 90% confidence level. To do this we find the standard error (as before):

$$SE = \frac{\sigma}{\sqrt{n}} = \frac{0.50 \text{ cm}}{\sqrt{100}} = \frac{0.50 \text{ cm}}{10} = 0.05 \text{ cm}$$

Then we use the *t* table (Table 9.1) to determine how many standard errors correspond to 90% confidence for a sample of 100. For $n = 100$, *d.f.* = 99, so we use the row for 120 *d.f.* The value in the column for 90% confidence is 1.658, which means that for a sample of this size an error range of 1.658 standard errors corresponds to a 90% confidence level. We thus multiply the standard error (0.05 cm) by 1.658 to arrive at an error range of ± 0.08 cm. We then say that we are 90% confident that our sample came from a population with a mean of 3.35 cm ± 0.08 cm. If our sample had consisted of 12 projectile points instead of 100, we would have had to use the row in the table for 11 *d.f.*, and we would have needed to use an error range of 1.796 standard errors instead of 1.658. Calibrating error ranges to a specific confidence level in this manner eliminates any possible confusion arising from differing sample sizes, and is generally to be recommended.

FINITE POPULATIONS

The example that we have used throughout this chapter involves a sample selected from a large and vaguely defined population—an infinite population in statistical terms. If the population is small and the sample is a substantial fraction of it, we can take mathematical advantage of an observation that makes intuitive good sense as well. It seems intuitively obvious that if our sample of 100 projectile points comes from a total population of 120 projectile points, then there is less uncertainty in our inference about the mean length in the population than if the sample of 100 comes from an effectively infinite population of projectile points. In this case at least, what seems true by common sense can be shown to be true mathematically as well. Whenever the population is finite we can include the *finite population corrector* in the equation for the standard error, thus:

$$SE = \frac{\sigma}{\sqrt{n}} \sqrt{1 - \frac{n}{N}}$$

where

σ = the standard deviation in the population (represented by the standard deviation in the sample as before);

n = the number of elements in the sample;

N = the number of elements in the population.

This will be recognized as the same equation used before for the standard error with the addition of the term ($\sqrt{1 - n/N}$). This term has a very simple effect. It makes the standard error smaller (hence the error range narrower and precision greater) if the sample is a very large portion of the population. For example, if we select a sample of 100 from a population of

120, $n = 100$, $N = 120$, and $\sqrt{1 - (n/N)} = \sqrt{1 - (100/120)} = 0.408$. Whatever the standard error would otherwise have been in such an instance, the addition of the finite population corrector makes it only .408 as large (multiplies it by 0.408). On the other hand, if the sample of 100 is selected from a population of 10,000, $n = 100$, $N = 10,000$, and $\sqrt{1 - (n/N)} = \sqrt{1 - (100/10,000)} = 0.99$. Multiplying whatever the standard error would otherwise have been by 0.99 clearly has very little effect on it.

The question arises, then, of when to apply the finite population corrector and when not to. It can always be applied when the number of elements in the population is known. If the population is very large compared to the size of the sample, it will not have much impact on the standard error. If you always use the finite population corrector when N is known, however, it will do its work whenever the sample is a large enough part of the population for it to make a difference. You cannot, of course, apply the finite population corrector when you do not know how many elements are in the population (that is, when the population is, for statistical purposes, infinite).

A COMPLETE EXAMPLE

The discussion of confidence levels and error ranges up to this point has made the whole process seem much more involved and complicated than it really is. This is a consequence of picking the process apart piece by piece to understand why it works the way it does. It is now time to work through an example without all the explanation to show that the procedure of inferring the mean of a population from a sample is really pretty straightforward.

Imagine that we have selected a random sample of 25 bowl rim sherds from the total of 53 bowl rim sherds recovered from a particular house in an excavated village site. We wish to infer the mean bowl rim diameter in the population of 53 rim sherds on the basis of measurements made on the 25 rim sherds in the sample, and we wish to state our inference at the 95% confidence level. The measurements are provided in Table 9.2. The stem-and-leaf plot in Table 9.2 confirms that the shape of this batch is roughly single-peaked and symmetrical (as least as much as it is reasonable to expect in a sample this small), so it seems reasonable to use the mean as an index of the level.

The mean of the 25 measurements is 14.79 cm, so the most likely single value for the mean rim diameter in the population of 53 rim sherds is 14.79 cm. The standard deviation in the sample is 3.21 cm, so the standard error is

$$SE = \frac{\sigma}{\sqrt{n}} \sqrt{1 - \frac{n}{N}}$$

$$= \frac{3.21 \text{ cm}}{\sqrt{25}} \sqrt{1 - \frac{25}{53}}$$

$$= \frac{3.21 \text{ cm}}{5} \sqrt{\frac{28}{53}}$$

$$= .64 \text{ cm} \sqrt{0.53}$$

$$= 0.47 \text{ cm}$$

Table 9.2. Rim Diameter Measurements for a Sample of 25 Rim Sherds

Diameter (cm)	Stem-and-leaf plot	
7.3		
9.3		
11.6		
11.8	21	0
12.2	20	
12.5	19	45
12.9	18	8
13.3	17	37
13.4	16	25
13.8	15	678
14.0	14	0489
14.4	13	348
14.8	12	259
14.9	11	68
15.6	10	
15.7	9	3
15.8	8	
16.2	7	3
16.5		
17.3		
17.7		
18.8		
19.4		
19.5		
21.0		
$\bar{X} = 14.79$ cm		
$\sigma = 3.21$ cm		

Since we need to state our inference at the 95% confidence level, we must find the value of t corresponding to the 95% confidence level and $n-1$ degrees of freedom. In the row of Table 9.1 for 24 *d.f.* and the column for 95% confidence, we find the t value 2.064. The error range we state, then, must be 2.064 standard errors. Since the standard error is 0.47 cm, the error range becomes $2.064 (0.47 \text{ cm}) = 0.97$ cm. We can thus state that we are 95% confident that the mean rim diameter for the 53 sherds recovered from this house is 14.79 cm $\pm$ 0.97 cm.

HOW LARGE A SAMPLE DO WE NEED?

If we know just what we need to find out before we select a sample, we are in position to determine how large a sample we need to study in order to achieve our objectives. We accomplish this by applying the same reasoning used throughout this chapter, but doing it backwards. That is, we decide in advance what confidence level we wish to speak at and how large an error range is acceptable. Then we determine how large a sample will be needed to meet those goals. The one quantity we must guess at is the likely magnitude of the standard deviation in the sample. Such a guess is often difficult to make in practice although it might be based on study of similar known samples.

For example, suppose we wish to estimate the mean thickness of sherds at a site with an error range no more than ± 0.5 mm at a confidence level of 95%. We have measured sherd thicknesses before for collections from a number of sites in the region and we find that the standard deviation in a sample of sherds is usually somewhere around 0.9 mm. We are willing to take the sherds visible on the surface to represent the sherds present in the site, and we want to send our field assistant to collect a sample of sherds randomly from the surface of the site. So as not to waste time, we would like to say in advance just how large a sample will be necessary. The error range (ER), of course, is t times the standard error, or

$$ER = t \left(\frac{\sigma}{\sqrt{n}} \right)$$

If we solve this formula for n, we get

$$n = \left(\frac{\sigma t}{ER} \right)^2$$

We have previously found the standard deviation in such samples to be about 0.9 mm, so we can use this value for σ. Since we do not yet know the sample size, we will use the row of Table 9.1 for ∞ d.f. to obtain a t value of 1.960 for a 95% confidence level. We want ER to be 0.5 mm. Thus

$$n = \left(\frac{(0.9 \text{ mm})(1.960)}{0.5 \text{ mm}} \right)^2$$

$$= \left(\frac{1.764 \text{ mm}}{0.5 \text{ mm}} \right)^2$$

$$= 3.528^2$$

$$= 12.447$$

We would tell our field assistant to select a sample of 12 or 13 sherds.

To show that this approach works, assume our field assistant returned with a sample of 13 sherds with a mean thickness of 7.3 mm and a standard deviation of 0.9 mm (as expected). The error range for a 95% confidence level would be

$$ER = t \left(\frac{\sigma}{\sqrt{n}} \right)$$

With a sample size of 13, we find that t for 12 d.f. and 95% confidence is 2.179, so

$$ER = 2.179 \left(\frac{0.9 \text{ mm}}{\sqrt{13}} \right)$$

$$= 2.179 \left(\frac{0.9 \text{ mm}}{3.606} \right)$$

$$= 2.179 \ (.250 \text{ mm})$$

$$= 0.54 \text{ mm}$$

Thus we would conclude that the mean thickness of sherds at the site in question is 7.3 mm ± 0.5 mm at the 95% confidence level. We have achieved our goal of estimating the thickness with an error range of no more than about 0.5 mm at the 95% confidence level.

Thinking about the confidence and precision we need in making specific estimates is one sound way to approach this always vexing question of

The Sample Size, the Sampling Fraction, and Rules of Thumb

The equations we have used in this chapter make clear that sample size is a very important issue. By sample size, statisticians ordinarily mean n, the number of elements in the sample. They do not nearly so often find it useful to think of sample size in terms of the sampling fraction (n/N, the fraction of the population included in the sample). They do not find it useful in the first place because so often samples are drawn from infinite populations (at least ones that are large and not enumerated). If we do not know how many elements are in the population we are sampling from, we clearly cannot even say what the sampling fraction is. In the second place, the number of elements in the sample clearly has much greater impact on the results of our calculations than the sampling fraction has. (If you do not believe this, try some experiments with the equations in this chapter, and you will see that it is true.)

This means' that when we begin to think about whether a sample is adequate for achieving our aims we must think less in terms of sampling fraction and more in terms of the number of elements in the sample. This clearly undermines one of the widespread misconceptions about sampling in archaeology. It has often been suggested that a good rule of thumb in sampling is to select a 5% sample. The principles discussed in this chapter make it quite clear that this is *not* a good rule of thumb. Sometimes a 5% sample will be insufficient, other times it will be far more than necessary; if the population is of undetermined size it will be inconceivable.

how large a sample is needed. Following this approach, of course, requires deciding specifically what we want to find out, how precise our results must be, and how confident we must be of our conclusions. These parameters are not absolutes. They vary from one situation to the next. What is sufficient precision in one context may be hopelessly imprecise in another. And what is sufficient confidence for some purposes may be altogether inadequate for others. If we cannot state our aims clearly enough at least to approximate how large a sample may be needed to achieve them, however, it is probably premature to be selecting a sample. We should go back and think harder about exactly what we are trying to find out.

ASSUMPTIONS AND ROBUST METHODS

The application of most of the techniques discussed in this volume requires making certain assumptions. These will be discussed at the close of each chapter. Most of the techniques are already fairly *robust*. That is, they can be applied to samples that only approximately meet the assumptions. And there are things to do with samples that even drastically violate the assumptions.

Once we have decided that we are willing to treat a batch of numbers as a random sample from a larger population we wish to know about, the only assumption we must make in order to estimate the population mean and attach error ranges to it in the manner described here is that the special batch must have an approximately normal distribution. The central limit theorem tells us that this will always be the case for large samples (that is, larger than 30 or 40 elements). When working with a smaller sample, it is wise to look at the stem-and-leaf plot to check for a roughly symmetrical and single-peaked shape. If a small sample has a single-peaked and roughly symmetrical shape, then we can count on its special batch to have a normal shape. If a small sample has a badly skewed shape we might try to correct this with transformations, but this is not very useful for estimating means because we would wind up estimating something like the mean of the logarithm of the population, and such a quantity is not very easy to relate to what we want to know.

Looking at a stem-and-leaf plot should always be the initial step anyway, even with a large sample. This is because the sample might have outliers or a badly skewed shape that would make the mean and standard deviation meaningless as numerical indexes of level and spread, as discussed in Chapters 2 and 3. If a sample has outliers or a badly skewed shape, then the population the sample was selected from probably does too. In such a case the trimmed mean and trimmed standard deviation are better indexes of the level and spread of the sample, also as discussed in Chapters 2 and 3. If so, then it makes sense to estimate not the regular mean of the population but the trimmed mean of the population instead. The best estimate of the trimmed mean of the population is simply the trimmed mean of the sample. Error ranges for different confidence levels can be provided for this estimate of the trimmed mean of the population following exactly the same procedures used to provide error ranges for estimates of the regular mean. The only difference is that, instead of using the sample size, the mean, and the standard deviation, their values are replaced in all equations with the trimmed sample size, the trimmed mean, and the trimmed standard deviation. Otherwise, everything about the calculations remains the same.

Table 9.3 lists a small sample of projectile point weights. The stem-and-leaf plot shows upward skewness and/or high outliers. The mean of this sample is 47.45 g, which falls too far above the center of the principal bunch of values to be a very useful index. If the sample is like this, the population probably is too. The trimmed mean would be a much more meaningful index of the center of such a shape. A 15% trimming fraction would eliminate the

**Table 9.3. Weights of a Small Sample of
Projectile Points**

Weight (g)	Stem-and-leaf plot	
96		
37	15	6
28	14	
34	13	
52	12	
18	11	
21	10	8
39	9	6
156	8	
43	7	
44	6	
19	5	25
30	4	347
108	3	014799
55	2	1488
24	1	89
28		
47		
39		
31		

three high outliers, which are causing most of the difficulty. The 15% trimmed mean, then, is 37.9 g, which falls where an index of the center of this batch should fall. The variance of the Winsorized batch is 137.19, so the trimmed standard deviation is 14.16. The standard error, then, is

$$SE = \frac{\sigma}{\sqrt{n}} = \frac{14.16}{\sqrt{14}} = 3.8\ g$$

For an error range at the 95% confidence level, we would multiply the standard error by the value of t for 13 d.f. $(n_T - 1)$. The 95% confidence error range, then, is ± 8.2 g (3.8×2.160). Estimating the trimmed mean instead of the regular mean for this population is not only more meaningful (it avoids the effects of the high outliers) but also more precise. The error range for 95% confidence that we would have to provide for an estimate of the regular mean would be ± 16.1 g. This is because the outliers that are eliminated by trimming would cause the standard deviation of the sample to be quite large. Consequently, the standard error and the 95% confidence error range would be quite large as well. Estimating the trimmed mean, then, pays off double in

---

Be Careful How You Say It

If you estimate the trimmed mean for a population rather than the regular mean, you must make it very clear what you've done. Be sure to refer to what you've estimated as the "trimmed mean," never just the "mean," and specify the trimming fraction as well. Just as with estimates of the regular mean, the confidence level for which the error range was calculated must be given too. For the example in the text, we might say, "On the basis of our sample, we estimate that the 15% trimmed mean weight of projectile points is 37.9 g ± 8.2 g at the 95% confidence level."

---

this instance—it is a more sensible index of the center here and its estimate comes with a much smaller error range.

PRACTICE

1. You have tested a newly reported neolithic site at Châteauneuf-sur-Loire. You decide that you are justified in working with the artifacts from your tests as if they were a random sample of the utilized flakes in the site. The lengths of the flakes are given in Table 9.4. Estimate an appropriate numerical index of center for length of utilized flakes in the site on the basis of this sample. Provide an error range for this estimate at the 95% confidence level. State in one clear sentence what this estimate and its error range mean.

2. You decide that the estimate you have made for utilized flake length at Châteauneuf-sur-Loire is not precise enough. You would like an estimate with an error range for 95% confidence that is only

Table 9.4. Lengths (in cm) of 40 Utilized Flakes from Châteauneuf-sur-Loire

4.7	6.8	3.5	5.9	6.5
4.1	6.2	6.0	7.8	8.8
8.0	9.3	8.3	8.1	7.4
3.2	6.9	5.5	4.3	8.5
9.7	7.3	4.3	4.7	6.3
7.5	4.5	4.8	3.0	7.0
5.7	3.9	5.6	6.1	5.3
5.0	5.4	6.1	5.1	2.6

Table 9.5. Diameters (in m) of 44 Mesolithic Hearths at Berwick-upon-Tweed

0.91	0.75	1.03	0.82	2.13
0.51	0.80	0.66	0.93	0.66
0.76	0.90	0.76	0.95	0.62
1.64	0.58	0.96	0.56	1.93
0.85	0.60	0.74	0.78	0.68
0.88	0.70	0.64	0.89	0.80
0.72	2.47	0.62	0.98	0.74
0.77	0.84	0.86	1.08	0.93
0.69	1.00	0.84	0.83	

Table 9.6. Zinc Content (in Parts Per Million) of 14 Obsidian Blades from a Prehistoric House at Huancabamba

53	49	41	59	74
37	66	33	48	57
60	55	82	22	

half as large as the one you just calculated, so you return to the site for more fieldwork in order to obtain a larger sample. How large a sample of utilized flakes will you need to achieve your aim?

3. You have excavated a mesolithic site at Berwick-upon-Tweed and found a remarkable number of well-formed hearths. Their diameters are given in Table 9.5. Using this set of hearths as a random sample of hearths at the site, estimate an appropriate numerical index of center for hearth diameters at the site as a whole. Provide an error range for this estimate at the 99% confidence level.

4. You have excavated the complete and well-preserved remains of a single prehistoric household at Huancabamba, and the artifacts recovered include 37 obsidian blades. In order to compare this assemblage with others and with different obsidian raw material sources, you wish to know the mean zinc content in the chemical composition of these 37 blades. Since zinc occurs in very small amounts, it is quite expensive to measure, so even though the entire assemblage is small, you treat it as a population from which you select a random sample of 14 blades to analyze. The quantity of zinc found in each blade (in parts per million) is given in Table 9.6. Estimate the mean number of parts per million of zinc in the population of 37 blades. Provide an error range for your estimate at the 90% confidence level. State the meaning of this estimate and its error range in a single clearly constructed sentence.

Chapter 10

Categories and Population Proportions

Chapters 7, 8, and 9 dealt with making inferences about a population on the basis of a sample when the observation of interest was a measurement whose mean in the population we wished to know. In Chapter 6 we discussed a different kind of observation, one based on categories rather than measurements. If the observation of interest involves a set of categories rather than a measurement, it of course makes no sense to think in terms of the center of a batch or its spread. Rather, we approach the batch in terms of proportions. When we observe categories in a sample, then, our basic thought about the population from which the sample was selected concerns the proportions of the different categories in the population, not the mean of anything.

The estimation of a population proportion on the basis of a sample is quite similar to the estimation of a population mean on the basis of a sample, so in this chapter we will treat proportions as an extension of the principles applied to means in the previous three chapters. Suppose that we examine the raw materials used to manufacture the projectile points in the sample of 100 projectile points discussed in Chapter 8. We may find that, of the 100 points, 13 are made of obsidian. Since the number in the sample is 100, the proportion of points made of obsidian in the sample is 13/100 or 13.0%. What does this tell us about the large and vaguely defined population that the sample of 100 points came from? Just as with means, the sample proportion is the likeliest single value for the proportion in the population from which the sample was selected. Thus, the best estimate of the population proportion, based on this sample, is 13.0%.

139

Just as it is possible that a sample may have a different mean than the population it came from, it is possible to select a sample with a proportion of 13.0% obsidian projectile points from a population with a proportion of obsidian projectile points different from 13.0%. Thus we would like to attach an error range and a confidence interval to this estimate just as we did to estimates of population means. We can use the standard error for this purpose in the case of proportions as well. The only difficulty is that calculation of the standard error of the mean was based on the standard deviation in the sample, and there is no obvious intuitive meaning to the concept of the standard deviation of a proportion. It can be shown mathematically, however, that there is a very simple equivalent of the standard deviation for proportions:

$$s = \sqrt{pq}$$

where

$s =$ the standard deviation of the proportion;
$p =$ the proportion expressed as a decimal fraction;
$q = 1 - p$.

In our example, the proportion of obsidian projectile points in the sample, expressed as a decimal fraction, is 0.130 and $q = 1 - p = 1 - 0.130 = 0.870$. Thus

$$s = \sqrt{pq} = \sqrt{(0.130)(0.870)} = \sqrt{0.1131} = 0.3363$$

This standard deviation is used in calculating the standard error by exactly the same procedure used for means. Since σ, the population standard deviation, is unknown, we use the sample standard deviation, s, in the equation

$$SE = \frac{\sigma}{\sqrt{n}} = \frac{0.3363}{\sqrt{100}} = \frac{0.3363}{10} = 0.03363$$

The standard error of the proportion in our example, then, is 0.034 or 3.4%. We can use this as a 1 standard error range attached to the estimated proportion and say that the population proportion is 13.0% ± 3.4%, or between 9.6% and 16.4%. As usual with a 1 standard error range, we would be about 66% confident that the proportion in the population our sample was selected from fell between 9.6% and 16.4%.

To adjust the error range thus obtained to the specifically desired confidence level, we would use Student's t distribution (Table 9.1) to determine t for the given number of degrees of freedom and confidence level and multiply the standard error by that value. To adjust the error range in this example to a 95% confidence level, we would use the row in Table 9.1 for 120 d.f. (the closest available to $n - 1 = 99$ d.f.), and find in the 95% confidence column that $t = 1.98$.

Multiplying the standard error by 1.98 yields (.034) (1.98) = .067. Thus, at a 95% level of confidence, we would estimate that the proportion of obsidian projectile points in the population from which our sample was selected is 13.0% ± 6.7% (or between 5.3% and 19.7%). This means, of course, that there is only a 5% chance of selecting a sample like ours (that is, a sample of 100 with a proportion of 13.0% obsidian projectile points) from a population with a proportion of obsidian projectile points less than 5.3% or greater than 19.7%).

The finite population corrector can be applied to the calculation of the standard error of a proportion just as with a mean. For example, suppose that in the complete excavation of a village site occupied for a relatively short period of time, we identify the remains of 24 houses. In the cases of 17 of the 24 houses, the remains are well enough preserved to enable us to determine the locations of the entrances. Of these 17 houses, 6 had their entrances facing south. After careful consideration of possible sources of bias, we decide that we will treat the 17 houses as a random sample from the population of 24 houses originally built at the site. We thus estimate that 6/17, or 35.3%, of the houses at the site had their entrances facing south. The standard error of this proportion will be

$$SE = \frac{\sigma}{\sqrt{n}} \sqrt{1 - \frac{n}{N}}$$

where $\sigma = s = \sqrt{pq}$.

Thus,

$$SE = \frac{\sqrt{pq}}{\sqrt{n}} \sqrt{1 - \frac{n}{N}} = \sqrt{\frac{pq}{n} \left(1 - \frac{n}{N}\right)}$$

$$= \sqrt{\left(\frac{(0.353)\,(0.647)}{17}\right)\left(1 - \frac{17}{24}\right)}$$

$$= \sqrt{(.0134)\,(1 - .7083)}$$

$$= 0.0625$$

If we wish to speak at a 90% confidence level, then we multiply this standard error by 1.746 (t for 90% confidence and 16 $d.f.$ is 1.746 according to Table 9.1) to get an error range at the 90% confidence level of 0.1091. We can thus conclude that of the 24 houses at the site, 35.3% ± 10.9% (or between 24.4% and 46.2%) had their entrances facing south. Since this is a finite population we can also convert this estimated proportion (and its attached error range) into numbers of houses for the entire population. Multiplying the lower extreme of the error range (24.4%) by the number of houses in the population (24) gives us 5.9 houses, and multiplying the upper extreme of the error range (46.2%) by the number of houses in the population (24) gives us 11.1 houses. Thus we can say that we are 90% confident that some 6 to 11 houses at the site had their entrances facing south.

In this example, the sample—and for that matter, the population from which it was selected—is so small that these statistical results may not seem very helpful. After all, we already knew there were at least 6 houses with their entrances facing south; there were 6 known south-facing entrances in the sample. And we knew there could not be more than 13 south-facing entrances. There were only 7 houses whose entrances were undocumented. If they all faced south, together with the 6 in the sample, that would make 13. If we already knew that the number of houses with south-facing entrances had to be between 6 and 13, what have we gained by saying that we have 90% confidence that the number of houses with south-facing entrances at this site is between 6 and 11? More than anything else, we have gained an awareness that our sample is quite small for saying anything very precise about the overall population with much confidence. For at least some purposes, this sample would simply be too small to tell us what we need to know, even though it was a 71% sample. (A sample of 17 houses represents 71% of the population of 24 houses.) A sample of 17 is, in statistical terms, a very small sample, no matter how large a proportion of the population it is. If we are working with a sample this small, there is an uncomfortably large risk that whatever proportions we find in it may be quite different from the proportions in the population from which it was selected. Whatever conclusions we derive from this sample about the population from which it was selected cannot be terribly precise or certain, even though they still do constitute our best guess about the population as a whole. Calculation of an error range for a specified confidence level, in this case, tells us that our best guess really is not very good, and *that* is important to know before we go on to use this observation as evidence for or against someone's theory.

HOW LARGE A SAMPLE DO WE NEED?

We can also put such knowledge to use in considering in advance roughly how large a sample we may need for a particular purpose, just as we can when estimating population means. The equation is the same:

$$n = \left(\frac{\sigma t}{ER}\right)^2$$

and we use $\sqrt{pq}$ for σ just as we did above. For example, suppose we wish to know how large a random sample of sherds we must collect from a site in order to estimate the proportions of the various pottery types in its ceramic assemblage with error ranges no wider than ± 5% at the 95% confidence level. We must make some guess at the proportions we may actually need to estimate in order to arrive at σ. If we have no idea, then we can use the most conservative guess of 50% since error ranges turn out to be widest when the proportion is 50%. To the extent that the actual proportions we get differ from 50%, then the error ranges will be even narrower than we require. Using 50%, $\sqrt{pq} = \sqrt{(0.50)(0.50)} = 0.50$, so we use 0.50 for σ. The value of t for 95% confidence and ∞ d.f. (since we do not yet know what n will be) is 1.96. Thus

$$n = \left(\frac{(.50)\,(1.96)}{.05}\right)^2 = 384.16$$

We should, then, collect a random sample of some 384 sherds.

If we do so, and discover that 192 of the sherds are of a particular ceramic type, then that type represents 192/384 or 50.0% of the sample. We estimate that the type composes 50.0% of the total ceramic assemblage at the site (the population from which the sample was selected). The standard error of this proportion is

$$SE = \frac{\sigma}{\sqrt{n}}$$

and we use $\sqrt{pq}$ for σ. Thus

$$SE = \frac{\sqrt{(0.50)(0.50)}}{\sqrt{384}} = \frac{0.50}{19.5959} = 0.0255$$

For an error range at the 95% confidence level, we multiply this standard error by the value of t for 95% confidence and ∞ d.f., since 383 d.f. falls far beyond 120, the last row of Table 9.1 before ∞. The 95% confidence level error range, then, is 1.96 standard errors: 0.050 or 5.0%. Thus we estimate that this ceramic type composes 50.0% ± 5.0% of the sherds at the site, and we have achieved the level of confidence and the precision that we required of our sample.

If another pottery type was represented by only 14 sherds in the sample, then we would estimate that it makes up 3.6% of the sherds at the site. In this case we would achieve greater precision at the same level of confidence because the standard error for this smaller proportion would be smaller:

$$SE = \frac{\sqrt{(0.036)(0.964)}}{\sqrt{384}} = \frac{0.1863}{19.5959} = 0.0095$$

Multiplying this standard error of 0.0095 by t for ∞ $d.f.$ and 95% confidence yields $(0.0095)(1.96) = 0.019$ or 1.9%. We could conclude with 95% confidence that this second pottery type represented 3.6% $\pm$ 1.9% of the ceramic assemblage at the site.

The difficulties of outliers and asymmetrical shapes that sometimes pose problems in the analysis of measurements and in the estimation of population means simply do not arise with categories and the estimation of population proportions. Thus it is not necessary to consider robust methods here.

PRACTICE

1. In systematic surface collection at the site of Mugombazi you recovered 342 flaked stone artifacts. After careful consideration of possible sources of sampling bias, you decide you will take these 342 as a random sample of the flaked stone in the site. Of the 342 flaked stone artifacts in the sample, 55 are identified as gravers. Estimate the proportion of gravers in the flaked stone assemblage at the site. Provide an error range for your estimate at the 99% confidence level. In one clearly constructed sentence, express this estimate, providing all the information your reader would need to know to make full use of it.

2. Not far south of Mugombazi lies another extremely large lithic scatter at Bwana Mkubwa. You intend to make a surface collection in such a manner as to have a random sample of the flaked stone at the site. Your aim is to estimate the proportions of different categories of flaked stone artifacts in the overall flaked stone assemblage, and you want estimates for which the error ranges (at a 90% confidence level) are never more than ±5%. How large a sample of artifacts should you select?

3. You proceed to Bwana Mkubwa and make the surface collection as planned (except, of course, for the incident with the rhinocerous).

To keep your mind off the pain, and to kill time while waiting in the emergency room, you have an initial look at the artifacts. It turns out that fully 45% of the flaked stone in the sample consists of debitage. Estimate the proportion of debitage in the flaked stone of the site as a whole. Provide an error range for your estimate at a 90% confidence level. State your results in a single clear sentence. Is the error range at least as small as the ±5% you wanted? If not, go back, figure out what went wrong, and try again. (This time, please watch out for the wildlife.)

Part III

Relationships between Two Variables

Chapter 11

Comparing Two Sample Means

Up to now we have concentrated on single batches of numbers and on using single batches of numbers as samples for purposes of making inferences about the populations from which they were selected. The principles discussed in Chapters 7–10, however, can also be applied to the task of comparing batches, which we began to explore in Chapter 4.

Figure 11.1 compares two batches of numbers. These batches are the areas (in meters) of house floors for two periods (Formative and Classic). After careful consideration of possible sources of bias we decide to work with these two batches as if they were random samples, taking each as a sample of house floors for its period. The sample for the Formative period consists of 32 house floors, and the sample for the Classic period consists of 52 house floors. We begin to explore the two samples with a back-to-back stem-and-leaf plot at the left in Figure 11.1. This plot reveals that both samples are single-peaked and symmetrical enough that the mean would be a useful index of their levels. Neither is a perfect single-peaked and symmetrical shape, but both are quite as single-peaked and symmetrical as one has any right to expect in relatively small batches of numbers like this.

The impression gained from the back-to-back stem-and-leaf plot is confirmed by the box-and-dot plot in the center of Figure 11.1. The box-and-dot plot in addition provides a clear view of the fact that the center of the sample of house floors from the Classic period is higher than the center of the

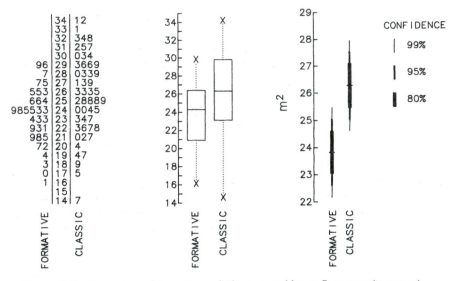

Figure 11.1. Comparison of Formative and Classic period house floor areas (in meters).

sample of floors from the Formative period. That is, Classic period houses, with a median area of 26.3 m^2, were in general somewhat larger than Formative period ones, with a median area of 24.3 m^2. (This remains a useful thing to say, even though there is considerable overlap in house size between the two periods and even though the smallest house of all dates to the Classic period—both of which facts are evident in the stem-and-leaf plot and in the box-and-dot plot.) The Classic period sample has a slightly larger spread than the Formative period sample does, although the two samples are not too different in this regard.

Table 11.1 provides the specific figures that compare the two samples in terms of level and spread. Whether we compare medians or means, Classic period house floors seem somewhat larger. And whether we compare mid-

Table 11.1. Comparison of Formative and Classic Period House Floor Samples

	Formative	Classic
$n =$	32 floors	52 floors
$Md =$	24.3 m^2	26.3 m^2
$\overline{X} =$	23.8 m^2	26.3 m^2
Midspread =	4.1 m^2	6.7 m^2
$s =$	3.4 m^2	4.5 m^2
SE =	0.60 m^2	0.63 m^2

spreads or standard deviations, Classic house floor areas show a slightly larger spread.

Combining these observations (of the sort we made in comparing batches in Chapter 4) with what we know of the behavior of random samples (see Chapters 7, 8, and 9) might lead us to wonder whether the differences we observe between these two batches are "real" or whether they are just the result of the simple fact that samples do not always very accurately represent the population from which they were selected. We know that if we selected a number of random samples from exactly the same population we would get considerable variation from one to the next. Such random variation between samples is often referred to as the *vagaries of sampling*. In comparing Formative and Classic period house floor areas, might we be seeing nothing more than this kind of random variation between samples? We know, of course, that these two samples did not actually come from the same population—one is a sample from Formative period house floors, and the other is from Classic period house floors. We often say that we imagine, though, that two such samples might have come from the same population, by which we really mean that the two samples might have come from two populations that have the same mean. If our two samples actually came from two populations with identical means, then the mean area of Formative and Classic house floors was the same. Since the means of our two samples are different, we would certainly guess that the means of the populations they were drawn from were different. Nevertheless, based on our previous discussion of the behavior of random samples, we recognize that there is some possibility that both samples could have been drawn from populations with means of, say, 25.0 m^2. If this were actually the case, then we would attribute the differences we observe in our samples of Formative and Classic period house floors to the vagaries of sampling. We would not take them to indicate a change in house floor area between the Formative and Classic periods. In Chapter 9 we dealt with this sort of question for one sample at a time by establishing error ranges for various confidence levels, but now we have two samples, which makes the situation more complicated. We can, nevertheless, approach the question in exactly the same way, taking each sample and its parent population in turn.

Table 11.2 provides estimates of Formative and Classic period house floor size for three different confidence levels. These estimates and their attached error ranges were calculated following exactly the procedures presented in Chapter 9. The two samples were treated independently, and error ranges for 80% confidence, 95% confidence, and 99% confidence were calculated separately on the basis of each sample. These error ranges are presented graphically at the right in Figure 11.1. Such graphs can be referred to as *bullet graphs* (because the representation of error ranges looks ever so slightly like a bullet). A bullet graph makes it easy to compare the two samples not

**Table 11.2. Estimates of Mean House Floor
Area for Different Confidence Levels**

	Mean area	
Confidence level	Formative	Classic
80%	$23.8 \pm .8 \text{ m}^2$	$26.3 \pm .8 \text{ m}^2$
95%	$23.8 \pm 1.2 \text{ m}^2$	$26.3 \pm 1.3 \text{ m}^2$
99%	$23.8 \pm 1.6 \text{ m}^2$	$26.3 \pm 1.7 \text{ m}^2$

only in terms of their means but also in terms of the confidence implications of their various error ranges. The thickest error bar represents the error range for the 80% confidence level. This is the most precise estimate and, correspondingly, the one in which our confidence is lowest. The medium thickness error bar represents the error range for the 95% confidence level. This error range is wider, but our confidence in this less precise estimate is higher. Finally the thinnest error bar represents the 99% confidence level error range, still less precise and thus worthy of still higher confidence. Note that this is not simply a different way of drawing a box-and-dot plot, although both bullet graph and box-and-dot plot represent centers and involve spreads. The box-and-dot plot in the middle of Figure 11.1 simply represents some characteristics of the two samples, while the bullet graph at the right represents some implications that the two samples have about the populations they were selected from. Note also that the scale of the bullet graph is different from the scale of the stem-and-leaf plot and the box-and-dot plot. Even the longest (99% confidence) error bars in the bullet graph are actually substantially shorter than the midspreads indicated in the box-and-dot plot. The scale is enlarged for the bullet graph so that the lengths of the error bars can be seen clearly and compared. If the bullet graph were drawn at the same scale as the box-and-dot plot, the error bars would be so short that they would not be as easy to see.

Comparing the two periods on the basis of the error bars at the right of Figure 11.1 yields the same results that the previous comparisons did with regard to level. Classic period house floors were larger on average than Formative period house floors. This graph, however, also helps us to answer the question about how likely it is that the differences between samples are nothing more than the random variation from one sample to the next that has to be expected even with no real difference between the populations from which the samples are selected.

We estimate that the mean house floor area during the Formative period was $23.8 \pm .8 \text{ m}^2$ at the 80% confidence level. That is, it is not very likely that our Formative sample came from a population with a mean less than 23.0 m^2 or greater than 24.6 m^2. Our estimate for Classic period house floors is a mean of 26.3 m^2. This is substantially outside the 80% confidence level error range for the Formative period. Thus there is less than a 20% chance that the Formative period sample came from a population with a mean as large as 26.3 m^2. The Classic period mean is also well outside the 95% confidence level error range for the Formative and even outside the 99% confidence level error range for the Formative. The error range for the Formative at the 99% confidence level reaches only to 25.4 m^2, still below the 26.3 m^2 mean for the Classic. Thus there is less than a 1% chance that the Formative period sample came from a population with a mean of 26.3 m^2. The probability, then, is less than 1% that we would get a sample like our Formative period one from a population like the Classic period population seems to be. By extension, there is less than a 1% chance that the difference we observe between Formative and Classic house floor areas in our samples could be a result of nothing more than the ordinary random variation from one sample to the next (samples, that is, drawn from the same population, which is to say from populations with the same mean). There is more than a 99% chance that we observe a difference in mean house floor area between our two samples because they really were drawn from populations that have different mean values. If our two samples are thus very likely to have been drawn from populations with different mean values, we know that it is very likely that the mean house floor area in the Formative really did differ from the mean house floor area in the Classic.

We would arrive at precisely the same conclusion if we made the comparison in the reverse direction by considering how likely it is that a sample like the Classic period one could be selected from a population of house floors like the Formative population seems to have been. The estimated mean for the Formative period falls well outside not only the 80% confidence level error range for the Classic period but also the 95% confidence level error range and the 99% confidence level error range as well. Thus the bullet graph in Figure 11.1 reveals very quickly to the eye that Formative period house floors are, on average, some 2.5 m^2 smaller than Classic period house floors and that this difference is very unlikely to be the result of the vagaries of sampling. That is, the graph tells us that the sizes and characteristics of the two samples upon which it is based are such as to give us considerable confidence (over 99% confidence) in saying that there was a change in house floor size from Formative to Classic period. A difference like the one we observe in our two samples would occur less than 1 time in 100 if the samples came from populations with the same mean. Not all statisticians are willing to take the next logical step of saying that this means there is less than 1 chance in 100 that Formative and Classic period house floors had the same mean area. Technically they are correct. The probabilities we are discussing are not actually probabilities about what the populations are like but rather probabilities about the samples we have. In practical terms, however, what any analyst does in a case like this is begin to talk with considerable confidence about a change in house floor area between the two periods.

CONFIDENCE, SIGNIFICANCE, AND STRENGTH

It would be more traditional statistical phrasing to say that the difference between Formative and Classic period house floor sizes is very significant. The statistical concept of significance is the mirror image of the concept of confidence as we have been using it. *Confidence* refers to the probability that the results we are stating *are not* attributable just to the vagaries of sampling. *Significance*, on the other hand, refers to the same concept from the opposite perspective—the probability that the results we are stating *are* attributable just to the vagaries of sampling. If we say, based on the comparison in Figure 11.1 and Table 11.2, that we find a difference between house floor areas in the Formative and Classic periods, the confidence we have in this statement is over 99%. The significance of the same statement is below 1%. The sum of the probability that corresponds to the level of confidence and the probability that corresponds to the level of significance is always 100%. Positive results in statistics, then, correspond to high confidence probabilities and, at the same time, to low significance probabilities. We are, to repeat, very confident that house floor areas are different in the two periods, which is the same as saying that the difference in house floor area between the two periods is very significant.

Both confidence and significance are concepts with quite clear and precise meanings in statistics (even if statisticians approach their definitions in many different ways). The notion of confidence in statistics corresponds pretty well to the colloquial use of the word "confidence." In common speech we say we are "confident" about something when we really do not think we are wrong. Paradoxically, however, the very act of saying that we are confident recognizes the possibility that we might be wrong at the same time that it classifies that possibility as a remote one. (If we really have no doubt at all about a fact, we usually just state it without even bothering to mention that we are quite confident of it.) The colloquial use of "significance," however, is rather different from its statistical use, and it is important not to confuse the two. We are likely to find something "significant," in colloquial speech, if it is important or meaningful. In statistics, however, "significant," like "confident," refers directly to the possibility that the conclusions we are stating are wrong—that is, the possibility that they represent nothing more than the normal variation to be expected in the random sampling process (that is, the vagaries of sampling).

The conclusion we arrived at in this example (that Classic houses were larger than Formative ones) may or may not be meaningful or important, but it *is* very significant. Whether it is meaningful or important is a substantive issue involved with what our interpretation of the result might be. The issue of meaningfulness or importance is an entirely separate one from that of confidence or significance. Staying purely in the realm of statistics, the closest we come to the issue of meaningfulness or importance is in the statistical concept of *strength*. In the comparison we have just made, the notion of strength is quite simple. The strength of the difference in house floor area between Formative and Classic is simply the magnitude of the difference, 2.5 m^2—the amount by which Formative period house floors appear to differ in area from Classic period ones on average.

We are highly confident in identifying this difference and we know that it is very significant—both statements meaning only that the difference we observe in our samples is not at all likely to be just the result of the vagaries of sampling. It is extremely likely that mean house floor size really was greater in the Classic than in the Formative. Whether this result is meaningful or important, however, has to do with why we are interested in this information in the first place. Perhaps we suspect a shift from nuclear family structure in the Formative to extended family structure in the Classic, and we reason that one way this might be evidenced in the archaeological record is in an increase in mean house floor area. We have found a very significant increase in mean house floor area, but it provides little support for our idea because the increase is too small (2.5 m^2) to be seen as an indicator of the need to provide more house space for substantially larger families. Both Formative and Classic period houses are, in general terms, relatively small even for nuclear family groups, and a change of only 2.5 m^2 is difficult to relate convincingly to a shift from households of perhaps four or five people to much larger households. Thus the result of our example investigation, while *highly significant*, was *not strong enough* to be important or meaningful, at least in this hypothetical interpretive context.

COMPARISON BY *t* TEST

The comparison between two samples that we have just made on the basis of bullet graphs of error ranges can also be approached as a significance test problem. This approach is entirely compatible with the comparison as we have already made it—it simply provides a different, and perhaps complementary, perspective on the situation. It also avoids one potential problem in making comparisons as we have just done. It is possible (with two samples of very different sizes, for example) for the comparison to yield different results depending on whether the error ranges from the first sample are compared to the mean of the second or vice versa. If the first sample is quite small, the error ranges may be quite large, leading to the conclusion that there is a very real possibility that such a sample could have come from a population like that indicated by the second sample. If the second sample is much larger, it may produce much smaller error ranges, leading to the contradictory conclusion that the second sample is very unlikely to have come from a population like that indicated by the first sample. The t test enables us to pool all the information from both samples into a single statement of the probability that both could be selected from the same population. Since in such situations we are likely to know that the two samples were, in fact, selected from different populations, this statement is really shorthand for saying "the probability that the two samples were selected from two populations with the same mean."

The two-sample t test evaluates the difference in means between the two samples in light of the pooled standard deviations from both samples. It is as if we calculated error ranges for the two based on the standard deviations from both together so that no matter whether we compared the first to the second or the second to the first, the results would be the same. The equation for accomplishing this at first seems formidable, but evaluating it is really quite a simple process of plugging in familiar values. First, the pooled standard deviation for the two samples is given by the expression

$$s_P = \sqrt{\frac{(n_1 - 1)s_1^2 + (n_2 - 1)s_2^2}{n_1 + n_2 - 2}}$$

where

s_P = the pooled standard deviation for the two samples;
n_1 = the number of elements in the first sample;
n_2 = the number of elements in the second sample;
s_1 = the standard deviation in the first sample;
s_2 = the standard deviation in the second sample.

Calculating this quantity for the Formative and Classic period house floor area samples used in the example above produces

$$s_P = \sqrt{\frac{(32-1)(3.4)^2 + (52-1)(4.5)^2}{(32 + 52 - 2)}}$$

$$= \sqrt{\frac{(358.36 + 1032.75)}{82}}$$

$$= \sqrt{(16.9648)}$$

$$= 4.12 \text{ m}^2$$

This pooled standard deviation for the two samples falls between the standard deviation of 3.4 m^2 for the Formative sample and the standard deviation of 4.5 m^2 for the Classic sample—which makes intuitive good sense. The pooled standard deviation is then the basis for a pooled standard error (SE_p):

$$SE_p = s_p \sqrt{\frac{1}{n_1} + \frac{1}{n_2}}$$

For the Formative and Classic house floor example,

$$SE_p = 4.12 \sqrt{\frac{1}{32} + \frac{1}{52}}$$

$$= 4.12 \sqrt{(.0505)}$$
$$= .93 \text{ m}^2$$

Knowing the pooled standard error enables us to say how many pooled standard errors the difference between sample means represents:

$$t = \frac{\bar{X}_1 - \bar{X}_2}{SE_p}$$

where

$\bar{X}_1$ = the mean in the first sample;
$\bar{X}_2$ = the mean in the second sample.

For the house floor area example,

$$t = \frac{23.8 - 26.3}{.93}$$
$$= -2.69$$

The observed difference in house floor area between the two samples, then, is 2.69 pooled standard errors. We know already that such a large number of standard errors is associated with high statistical confidence, and

Be Careful How You Say It

When presenting the result of a significance test, it is always necessary to say just what significance test was used and to provide the resulting statistic and the associated probability. For the example in the text, we might say, "The 2.5 m^2 difference in mean house floor area between the Formative and Classic periods is very significant ($t = -2.69$, $.01 > p > .005$)." This one sentence really says everything that needs to be said. No further explanation would be necessary if we were writing for a professional audience whom we can assume to be familiar with basic statistical principles and practice. The "statistic" in this case is t, and providing its value makes it clear that significance was evaluated with a t test, which is quite a standard technique that does not need to be explained anew each time it is used. The probability that the observed difference between the two samples was just a consequence of the vagaries of sampling is the *significance*, or the *associated probability*. Ordinarily p stands for this probability, so in this case we have provided the information that the significance is less than 1%. This means the same thing as saying that our confidence in reporting a difference between the two periods is greater than 99%.

 If, instead of performing a t test, we simply used the bullet graph to compare estimates of the mean and their error ranges, as in Figure 11.1, we might say, "As Figure 11.1 shows, we can have greater than 99% confidence that mean house floor area changed between Formative and Classic periods." The notion of estimates and their error ranges for different confidence levels is also a very standard one which we do not need to explain every time we use it. Bullet graphs, however, are less common than, say, box-and-dot plots, so we cannot assume that everyone will automatically understand the specific confidence levels of the different widths of the error bars. A key indicating what the confidence levels are, as in Figure 11.1, is necessary.

 In an instance like the example in the text, a bullet graph and a t test are alternative approaches. Using and presenting both in a report qualifies as statistical overkill. Pick the one approach that makes the simplest, clearest, most relevant statement of what needs to be said in the context in which you are writing; use it; and go on. Presentation of statistical results should support the argument you are making, not interrupt it. The simplest, most straightforward presentation that provides complete information is the best.

thus with the low probability values that mean great significance as well. To be more specific, this t value can be looked up in Table 9.1. The number of degrees of freedom is $n_1 + n_2 - 2$, or in this example $32 + 52 - 2 = 82$. Thus we use the row for 60 *d.f.*, which is the closest row to 82. Ignoring the sign for the moment, we look for 2.69 in this row. It would fall between the columns for 1% and 0.5% significance. Thus the probability that the difference we observe between the two samples is just due to the vagaries of sampling is less than 1% and greater than 0.5%. We could also say that the probability of selecting two samples that differ as much as these do from populations with the same mean value is less than 1%. Yet another way to express the same thought is that we are more than 99% confident that average house floor areas differed between the Formative and Classic periods. This is, of course, the same conclusion that was already apparent in Figure 11.1 and that we have discussed earlier.

The sign of the t value arrived at indicates the direction of the difference. If the second population has a lower mean than the first, t will be positive. If the second population has a higher mean than the first, t will be negative. The strength of the difference is still indicated simply by the difference in means between the two samples, as it has been all along: 2.5 m^2.

THE ONE-SAMPLE t TEST

Occasionally we are interested in comparing a sample not to another sample but to some particular theoretical expectation. For example, we might be interested in investigating whether a particular prehistoric group practiced female infanticide. One line of evidence we might pursue would be sex ratios in burials. Suppose we had a sample of 46 burials, which we were willing to take as a random sample of this prehistoric population except for infants intentionally killed, whose bodies we think were disposed of in some other way. On theoretical grounds we would expect this sample of burials to be 50% males and 50% females, unless sex ratios were altered by some practice such as female infanticide. (Actually there might be reason to expect very slightly different proportions from 50:50 on theoretical grounds, but that does not really affect what concerns us here.) After careful study of the skeletal remains, we determine that 21 of the 46 burials were females and 25 were males. The proportions are thus 45.7% female and 54.3% male. This lower proportion of females in our sample might make us think that more females were killed in infancy than males, but we wonder how likely it is that we could select a random sample of 46 with these proportions from a population with an even sex ratio. We could calculate the error ranges for various levels of confidence, as we did in Chapter 10. For a proportion of 45.7% females, in a sample of 46, the standard error would be

$$SE = \frac{\sqrt{pq}}{\sqrt{n}} = \frac{\sqrt{(.457)(.543)}}{\sqrt{46}} = \frac{.498}{6.782} = .073$$

For, say, an 80% level of confidence, we would look up the value of t for 80% confidence and 45 degrees of freedom. We would multiply the standard error by this t value: $(.073)(1.303) = .095$. We would thus be 80% confident that our sample of 46 burials was drawn from a population with 45.7% ± 9.5% females (or between 36.2% and 55.2% females). The theoretical expectation of 50% females is well within this 80% confidence level error range, so there is something over a 20% chance that the divergence from even sex ratios that we observe in our sample is only the result of sampling vagaries.

To be more precise about it with a one-sample t test, we would simply use the standard error and the t table in a slightly different way. The observed proportion of females in our sample is 45.7%, or 4.3% different from the expected 50:50 ratio. This difference of .043 (that is, 4.3%) represents .589 standard errors since $.043/.073 = .589$. Looking for this number of standard errors on the row of the t table corresponding to 40 degrees of freedom (the closest we can get to 45 degrees of freedom) would put us slightly to the left of the first column in the body of the table (the one that corresponds to 50% significance). There is something over a 50% chance, then, of getting a random sample of 46 with as uneven a sex ratio as this from a population with an even sex ratio. This means that it is uncomfortably likely that the uneven sex ratios we observe in our sample are nothing more than the vagaries of sampling. We might also say, "The difference we observe between our sample and the expected even sex ratio has extremely little significance ($t = .589$, $p > .5$)." These results would not provide much support for the idea of female infanticide. At the same time they would not provide much support to argue that female infanticide was *not* practiced since there is also an uncomfortably large chance that this sample could have come from a population with an uneven sex ratio. In short, given the proportions observed, this sample is simply not large enough to enable us to say with much confidence whether the population it came from had an even sex ratio or not.

THE NULL HYPOTHESIS

Significance tests are often approached by practitioners of many disciplines as a question of testing hypotheses. In this approach, first a *null hypothesis* is framed. In the example of Formative and Classic house floor areas, the null hypothesis would postulate that the observed difference between the two samples was a consequence of the vagaries of sampling. An arbitrary significance level would be chosen for rejecting this hypothesis. (The level chosen is almost always 5%, for no particularly good reason.) And then the t test would be performed. The result ($t = -2.69$, $.01 > p > .005$) is a significance level that exceeds the usual 5% rejection level. (That is, the probability that the difference is just due to the vagaries of sampling is even less than the chosen 5% threshold.) Thus the null hypothesis (that the difference is just random sampling variation) is rejected, and the two populations are taken to have different mean areas.

The effect of framing significance tests in this way is to provide a clear yes or no answer to the question of whether the observation tested really characterizes the populations involved rather than just being the result of sampling vagaries. The problem is that statistics never really do give us a yes or no answer to this question. Significance tests may tell us that the probability that the observation is just the result of sampling vagaries is very high or moderate or very low. But as long as we are making inferences from samples we are never absolutely certain about the populations the samples represent. Significance is simply not a condition that either exists or does not exist. Statistical results are either more significant or less significant. We have either greater or lesser confidence in our conclusions about populations, but we *never have absolute certainty*. To force ourselves either to reject or to accept a null hypothesis is to oversimplify a more complicated situation to a yes or no answer. (Actually, many statistics books make the labored distinction that one does not accept the null hypothesis but rather "fails to reject" it. In practice, analysts often treat a null hypothesis they have been unable to reject as a proven truth—more on this subject later.)

This practice, of forcing statistical results like "maybe" and "probably" to become "no," and "highly likely" to become "yes," has its clearest justification in areas like quality control, where an unequivocal yes or no decision must be made on the basis of significance tests. If a complex machine turns out some product, a quality control engineer may test a sample of the output to determine whether the machine needs to be adjusted. On the basis of the sample results, the engineer must decide either to let the machine run (and risk turning out many defective products if he or she is wrong) or stop the machine for adjustment (and risk wasting much time and money if he or she is wrong). In such a case statistical results like "the machine is probably turning out defective products" must be converted into a yes or no answer to the question of stopping the machine. Fortunately, research archaeologists are rarely in such a position. We can usually (and more informatively) say things like "possibly," "probably," "very likely," and "with great probability."

Finally, following the traditional 5% significance rule for rejecting the null hypothesis leaves us failing to reject the null hypothesis when the probability that our results are just the vagaries of sampling is only 6%. If, in the house floor example, the t value had been lower, and the associated probability had been 6%, it would have been quite reasonable for us to say, "We have fairly high confidence that mean house floor area was greater in the Classic period than in the Formative." If we had approached the problem as one of attempting to reject a null hypothesis, however, with a 5% rejection level, we would have been forced to say instead, "We have failed to reject the hypothesis that house floor areas in the Formative and Classic are the same." As a consequence we would probably have proceeded as if there were no difference in house floor area between the two periods when our own statistical results had just told us that there was a 94% probability that there *was* such a difference.

In some disciplines, almost but not quite rejecting the null hypothesis at the sacred 5% level is dealt with by simply returning to the lab or wherever and studying a larger sample. Other things being equal, larger samples produce higher confidence levels, and higher confidence levels equate to lower significance probabilities. Almost but not quite rejecting the null hypothesis, then, can translate into, "There is probably a difference, but the sample was not large enough to make us as confident as we would like about it." In archaeology, however, it is often difficult or impossible to simply go get a larger sample, so we need to get all the information we can from the samples we have. For this reason, in this book we will approach significance testing not as an effort to reject a null hypothesis but instead as an effort to say just how likely it is that the result we observe is attributable entirely to the vagaries of sampling.

Table 11.3 summarizes the differences between a null hypothesis testing approach to significance testing and the more scalar approach advocated in this book. The approach followed here can, of course, be thought of as testing the null hypothesis but not forcing the results into a yes or no decision about it. It we are willing to take a more scalar approach, though, there is no advantage to plunging into the confusion of null hypothesis formulation, rejection, and failure to reject. In particular, Table 11.3 emphasizes how potentially misleading is the answer "true" when applied to a full range of probabilities concerning the null hypothesis that can more accurately be described as ranging from "extremely likely" to "fairly unlikely."

Pregnancy tests have only two possible results: pregnant and not pregnant. Significance tests are simply not like that; their results run along a continuous scale of variation from very high significance to very low significance. While some users of statistics (like gamblers) find themselves having to answer yes or no questions on the basis of the probabilities given by significance tests, archaeologists can count themselves lucky that they are not often in such a situation. We are almost always able to say that results provide very strong support for our ideas, or moderately strong support, or some support, or very little support. Forcing significance tests simply to reject or fail to reject a null hypothesis, then, is usually unnecessary and unhelpful in archaeology and may do outright damage by being misleading as well. In this book we will never characterize results as simply "significant" or "not significant" but rather as more or less significant with descriptive terms akin to those in Table 11.3. Some statistics books classify this proce-

Table 11.3. Summary of Contrasting Approaches to Significance Testing in the Context of the House Floor Area Example

Significance testing as an effort to reject the null hypothesis (not recommended here).	Significance testing as an effort to evaluate the probability that our results are just the vagaries of sampling (the approach followed in this book).

The questions asked:

The difference observed between the Formative and Classic house floor samples is nothing more than the vagaries of sampling. True or false?	How likely is it that the difference observed between Formative and Classic house floor samples is nothing more than the vagaries of sampling?

Example answers for different possible significance levels:

$p = .80$	True.	Extremely likely.
$p = .50$	True.	Very likely.
$p = .20$	True.	Fairly likely.
$p = .10$	True.	Not very likely.
$p = .06$	True.	Fairly unlikely.
$p = .05$	False.	Fairly unlikely.
$p = .01$	False.	Very unlikely.
$p = .001$	False.	Extremely unlikely.

dure as a cardinal sin. Others find it the only sensible thing to do. The fact is that neither approach is truth revealed directly by God. Archaeologists must decide which approach best suits their needs by understanding the underlying principles, not by judging which statistical expert seems most Godlike in revealing his or her "truth."

STATISTICAL RESULTS AND INTERPRETATIONS

It is easy to accidentally extend levels of confidence or significance probabilities beyond the realm to which they properly apply. Either one is a statistical result that takes on real meaning or importance for us only when interpreted. In the example of Formative and Classic period house floors used throughout this chapter, our interest, as suggested earlier, may be investigating a possible shift from nuclear families in the Formative to extended families in the Classic. Given the samples we have in this example, we find that houses in the Classic were larger, on average, than houses in the Formative. We also find that this difference is very significant (or that we have quite high confidence that it is not just the result of sampling vagaries, which means the same thing). This does not, however, automatically mean that we have quite high confidence that nuclear family organization in the Formative changed to extended family organization in the Classic. The former is a statistical result; the latter is an interpretation. How confident we are in this latter interpretation depends on a number of things in addition to the statistical result. For one thing, already mentioned early in this chapter, despite the high significance of the size difference between Formative and Classic house floors, the strength of the difference (2.5 m²) is not very much—at least not compared to what we would expect from such a change in family structure. There might be several other completely different kinds of evidence we could bring to bear as well. It would only be after weighing all the relevant evidence (among which our house floor area statistical results would be only one item) that we would be prepared to assess how much confidence we have in the suggested interpretation. We would not really be in position to put a number on our confidence in the interpretation about family structure, because it is an interpretation of the evidence, not a statistical result. We would presumably need to weigh the family structure interpretation against other possible interpretations of the evidence. While statistical evaluation of the various lines of evidence is extremely helpful in this process, its help comes from evaluating the confidence we should place in certain patterns observable in the measurements we make on the samples we have, not from placing probabilities directly on the interpretations themselves. These interpretations are connected sometimes by a very long chain of more or less convincing logical links or assumptions to the statistical results obtained by analyzing our samples.

ASSUMPTIONS AND ROBUST METHODS

The two-sample t test assumes that both samples have approximately normal shapes and roughly similar spreads. If the samples are large (larger than 30 or 40 elements), violations of the first assumption can be tolerated because the t test is fairly robust. As long as examination of a box-and-dot plot reveals that the midspread of one sample is no more than twice the midspread of the other, the second assumption can be considered met. If the spreads of the two samples are more different than this, then that fact alone suggests that the populations they came from are different, and that, after all, is what the two-sample t test is trying to evaluate.

If the samples to be compared contain outliers, the two-sample t test may be very misleading, based as it is on means and standard deviations, which will be strongly affected by the outliers, as discussed in Chapters 2 and 3. In such a case an appropriate approach is to base the t test on the trimmed means and trimmed standard deviations, also discussed in Chapters 2 and 3. The calculations for the t test in this case are exactly as they are for the regular t test except that the trimmed sample sizes, the trimmed means, and the trimmed standard deviations are used in place of the regular sample sizes, the regular means, and the regular standard deviations.

If the samples to be compared are small and have badly asymmetrical shapes, this can be corrected with transformations, as discussed in Chapter 5, before performing the t test. The data for both samples are simply transformed and the t test is performed exactly as described above on the transformed batches of numbers. It is, of course, necessary to perform the same transformation on both samples, and this may require a compromise decision about which transformation produces the most symmetrical shape simultaneously for both samples.

PRACTICE

You have just completed extensive excavations at the Ollantaytambo site. You select a random sample of 36 obsidian artifacts from those recovered at the site for trace element analysis in an effort to determine the source(s) of the raw material. You realize that you really should investigate a whole suite of elements, but the geochemist you collaborate with only gives you the data for the element zirconium before he sets off down the Urubamba River in a dugout canoe, carrying the remainder of the funds you have budgeted for raw material sourcing. There are two visually different kinds of obsidian in the sample—an opaque black obsidian and a streaky gray obsidian—and you know that such visual distinctions sometimes correspond to different sources. The data on amounts of zirconium and color for your sample of 36 artifacts are given in Table 11.4.

Table 11.4. Zirconium Content for a Sample of Gray and Black Obsidian Artifacts from Ollantaytambo

Zirconium content (ppm)	Color	Zirconium content (ppm)	Color	Zirconium content (ppm)	Color
137.6	Black	136.4	Black	138.6	Gray
135.3	Gray	138.8	Black	138.6	Black
137.3	Black	136.8	Gray	139.0	Black
137.1	Gray	136.3	Gray	131.5	Gray
138.9	Gray	135.1	Black	142.5	Black
138.5	Gray	132.9	Gray	137.4	Gray
137.0	Gray	136.2	Gray	141.7	Black
138.2	Black	139.7	Gray	136.0	Gray
138.4	Black	139.1	Black	136.9	Black
135.8	Gray	139.2	Gray	135.0	Gray
137.4	Black	132.6	Gray	140.3	Black
140.9	Black	134.3	Gray	135.7	Black

1. Begin to explore this sample batch of zirconium measurements with a back-to-back stem-and-leaf diagram to compare the black and gray obsidian. What does this suggest about the source(s) from which black and gray obsidian came?

2. Estimate the mean measurement for zirconium for gray obsidian and for black obsidian in the populations from which these samples came. Find error ranges for these estimates at 80%, 95%, and 99% confidence levels, and construct a bullet graph to compare black and gray obsidian. How likely does this graph make it seem that gray and black obsidian came from a single source?

3. What, exactly, have you calculated the probabilities of in Question 2? What, exactly, are the populations you have characterized? What are the logical links necessary to use this evidence in support of the conclusion you want to make about obsidian sources?

4. Approach Question 2 using a t test. State the conclusion derived from your t test in a single clear sentence as if you were reporting it in a paper.

Chapter *12*

Comparing Means of More Than Two Samples

In Chapter 11 we took two approaches to comparing the means of two samples. The first approach involved using each sample separately to estimate the mean of the population that the sample came from. We then attached error ranges for several confidence levels to these estimates and drew a picture of the whole thing with a bullet graph (Fig. 11.1). This approach is easily extended to the comparison of any number of samples. In this chapter we will use another fictitious example consisting of 127 Archaic period projectile points from the Cottonwood River valley. After considering possible sources of bias, we decide to work with these as a random sample from the large and vaguely defined population of Archaic projectile points from the Cottonwood River valley.

We are interested in whether, during the Archaic period, there was much change in hunting of large and small animals in the Cottonwood River valley. We reason that large projectile points are more involved in hunting large animals and small projectile points are more involved in hunting small animals. We can divide the 127 projectile points into three groups: Early, Middle, and Late Archaic, and we decide to compare the weights of projectile points in these three periods. One way to organize these data for this sample is shown in Table 12.1. Here two observations are recorded for each of the

Table 12.1. Data on Weight and Period for a Sample of Archaic Period Projectile Points from the Cottonwood River Valley

Weight (g)	Archaic subperiod	Weight (g)	Archaic subperiod	Weight (g)	Archaic subperiod
54	Early	30	Early	63	Middle
39	Early	52	Early	64	Middle
49	Early	56	Early	78	Middle
65	Early	63	Early	62	Middle
54	Early	53	Early	78	Middle
83	Early	79	Early	57	Middle
75	Early	50	Early	59	Middle
45	Early	54	Early	31	Middle
68	Early	51	Early	69	Middle
47	Early	59	Early	32	Middle
57	Early	60	Early	69	Middle
19	Early	48	Early	80	Middle
47	Early	40	Early	78	Middle
58	Early	50	Early	69	Middle
76	Early	69	Early	34	Late
50	Early	71	Middle	39	Late
67	Early	64	Middle	40	Late
52	Early	59	Middle	45	Late
40	Early	65	Middle	37	Late
58	Early	54	Middle	32	Late
42	Early	65	Middle	31	Late
43	Early	63	Middle	60	Late
58	Early	52	Middle	58	Late
28	Early	44	Middle	45	Late
59	Early	73	Middle	50	Late
43	Early	70	Middle	40	Late
45	Early	56	Middle	41	Late
60	Early	46	Middle	38	Late
27	Early	61	Middle	59	Late
64	Early	49	Middle	37	Late
73	Early	51	Middle	28	Late
70	Early	61	Middle	37	Late
68	Early	70	Middle	31	Late
68	Early	51	Middle	40	Late
85	Early	42	Middle	34	Late
49	Early	73	Middle	37	Late
21	Early	51	Middle	44	Late
24	Early	74	Middle	47	Late
50	Early	40	Middle	54	Late
52	Early	67	Middle	36	Late
62	Early	51	Middle	48	Late
44	Early	59	Middle		
61	Early	68	Middle		

127 projectile points: the weight (in grams) and the period (Early, Middle, or Late Archaic). Our two variables, weight and period, are of different kinds. Weight, of course, is a measurement, and period is a set of three categories.

COMPARISON WITH ESTIMATED MEANS AND ERROR RANGES

We can use the three period categories to separate the sample of 127 projectile points into three samples—one consisting of the 58 early Archaic points, one of the 42 Middle Archaic points, and one of the 27 Late Archaic points. If we were willing to treat the 127 projectile points as a random sample from the Archaic projectile points of the Cottonwood River valley, then we can be equally willing to treat the 58 Early Archaic points as a random sample of the Early Archaic points of the Cottonwood River valley, the 42 Middle Archaic points as a random sample of the Middle Archaic projectile points of the Cottonwood River valley, and the 27 Late Archaic points as a random sample of the Late Archaic points of the Cottonwood River valley. If we do this, then we have reorganized a single batch of numbers into three batches of numbers that can be compared just as we compared the two batches of numbers in Chapter 11.

Table 12.2 provides numerical indexes (sample size, mean, standard deviation, standard error, and variance) for each of these three smaller samples. Using the standard errors and Table 9.1 we can provide estimated mean weights for each of the three populations these samples came from and present the whole comparison graphically as in Figure 12.1. The Early Archaic and Middle Archaic samples are large enough that we can count on the special batch having a normal shape. The late Archaic sample is a bit small for us to count on a normal shape for the special batch, so we look at the stem-and-leaf plot for Late Archaic in Figure 12.1 to make sure that the sample itself has a fairly normal shape (which it does). The box-and-dot plot makes clear that Middle Archaic projectile points tend to be the heaviest and Late Archaic projectile points the lightest, with Early Archaic projectile point

Table 12.2. Comparison of Weights of Projectile Points for Archaic Subperiods

	Early	Middle	Late	All Archaic
$n =$	58	42	27	127
$\bar{X} =$	53.67 g	60.45 g	41.56 g	53.34 g
$s =$	14.67 g	12.15 g	8.76 g	14.42 g
$SE =$	1.93 g	1.88 g	1.69 g	1.28 g
$s^2 =$	215.21	147.62	76.74	207.94

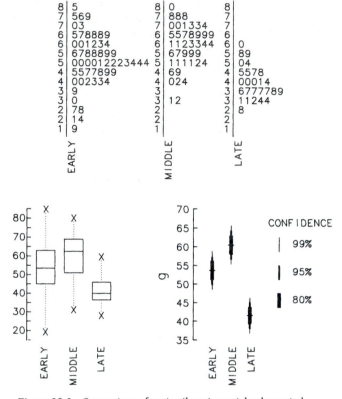

Figure 12.1. Comparison of projectile point weights by period.

weights falling in between. The ranges of all three certainly overlap, however, especially those of Early Archaic and Middle Archaic. At the far right in Figure 12.1, the bullet graph of estimated population means with error bars for confidence levels of 80%, 95%, and 99% makes it clear that the differences between the three samples we have are very highly significant. None of the error ranges for 99% confidence includes the estimated mean of any of the other populations. We are thus more than 99% confident that the differences we observe between samples are not just a consequence of the vagaries of sampling. It is extremely likely instead that such different samples came from parent populations that differed from each other.

Figure 12.1 demonstrates once again that box-and-dot plots and bullet graphs are two different things. The boxes representing the midspreads for the three periods overlap substantially, while the error ranges for 80%, 95%, and 99% confidence do not. Since these two kinds of plots are similar in appearance and since both deal with the spreads of batches in one way or another, it is easy to overlook the fundamental difference between the two. While it is true that the error ranges in the bullet graph in Figure 12.1 are based, in part, on the spread of each batch, they are not simply a graphical representation of that spread. They rely as well on the sample sizes and are thus a picture not of the spreads of the three sample batches but rather of the spreads of the corresponding special batches, as discussed in Chapter 8. As a consequence, the bullet graphs have useful implications concerning the parent populations that the box-and-dot plots do not have.

COMPARISON BY ANALYSIS OF VARIANCE

Estimating means and attaching error ranges to each estimate provides a good way to compare each sample with each other sample, and a bullet graph literally draws the overall picture. In terms of significance, this overall picture is summed up in the question, "How likely is it that we could get three samples with means and standard deviations like these from a single parent population?" Another way to say it would be, "What is the probability that samples as different as these three could be produced from the same population just through the vagaries of sampling?" When we speak of a "single parent population" or the "same population" in these questions, we are speaking metaphorically, since we know the three samples came from three different populations—one of Early Archaic projectile points, one of Middle Archaic projectile points, and one of Late Archaic projectile points. Hypothesizing here that they may have come from the "same population" is simply shorthand for inquiring how likely it is that the three populations these three samples came from had the same mean. Thus, the significance question we are asking amounts to, "How likely is it that Early Archaic, Middle Archaic, and Late Archaic projectile point populations all had the same mean weight, and that our three samples differ just because random samples, even from the same population, do differ from each other?"

We answered such a question with a two-sample t test in Chapter 11, but this test cannot easily be extended to more than two samples. For three or more samples, the technique of choice is *analysis of variance*, often abbreviated *ANOVA*. As the name implies, analysis of variance relies on variance as the key to answering the significance question in this situation. (Remember that the variance of a batch is simply the square of the standard deviation.) The variances (s^2) of all three separate subsamples and of the entire sample of 127 are given in Table 12.2.

Analysis of variance assumes that the samples are drawn from populations with normal shapes. We examine the stem-and-leaf plots for the three separate subsamples in Figure 12.2, and we see the fundamentally single-

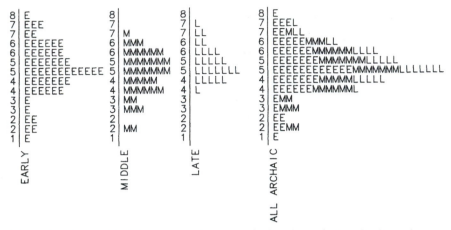

Figure 12.2. Stem-and-leaf plots of projectile point weights by subperiod where all subperiod groups have similar means.

peaked and symmetrical shape that we need to see for each of the subsamples. Analysis of variance also assumes that the spreads (specifically the variances) of the populations are approximately equal. The box-and-dot plots in Figure 12.1 provide an easy way to judge the spreads of the samples, as do the figures for the variances given in Table 12.2. Here the largest variance is almost 3 times as big as the smallest. Comparing midspreads in the box-and-dot plots yields a similar observation. This difference in spreads is pressing the limits of analysis of variance's ability to withstand violation of its basic assumptions. As long as the largest variance is no more than 3 times the smallest, though, we are willing to go ahead and perform analysis of variance, especially if the samples involved are not too small.

Figure 12.2 illustrates one possible result of a comparison of weights for the three subsamples of projectile points from different parts of the Archaic period. Note that Figure 12.2 does not really illustrate the data presented in Table 12.1. Instead, it illustrates one pattern that we *might* have seen. This pattern has been created by maintaining the real shapes of all three subsamples but shifting their centers so that they fall much closer together for purposes of discussion only. The stem-and-leaf plots in Figure 12.2 are drawn with letters standing for the different subperiods in order to make it possible to see what happens when the three subsamples are combined, as at the extreme right.

When we compare the overall sample of 127 projectile points in Figure 12.2 to the individual subsamples, we observe several things. First, in this result, all three subsamples look pretty much the same. All three have centers

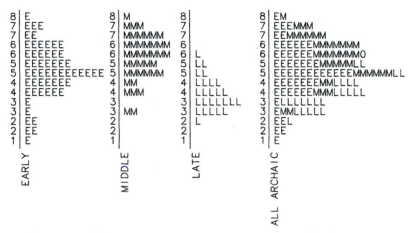

Figure 12.3. Stem-and-leaf plots of projectile point weights by subperiod for the data presented in Table 12.1.

in about the same place. All three have roughly similar spreads. Second, the spread of the overall sample of 127 projectile points is similar to the spreads of the individual subsamples. And third, the center of the overall sample of 127 projectile points is quite similar to the centers of the individual subsamples. Despite some minor differences in shape, all four stem-and-leaf plots are fairly similar. The sharpest difference is that the peak in the stem-and-leaf for the overall sample is considerably higher than the peaks for the individual subsamples. This should not be surprising, since the overall sample has considerably more projectile points, but a spread not really larger than those of the individual subsamples. Consequently they mount up higher at the peak.

A different possible result of such a comparison is illustrated in Figure 12.3, and this figure *does*, in fact, accurately reflect the data in Table 12.1. Comparing Figure 12.3 with Figure 12.2 reveals the nature of the differences. First, the three subsamples no longer look pretty much the same. Their spreads continue to be roughly similar, but their centers are clearly in different places. Second, the spread of the overall sample is larger in Figure 12.3 than in Figure 12.2. It is no longer as close to the spreads of the individual subsamples as it was in Figure 12.2. While the Early Archaic subsample has the largest spread, and this continues to be comparable to the spread in the overall sample, the Middle Archaic and Late Archaic subsamples have noticeably narrower spreads than the overall batch. And third, the center of the overall sample, while similar to the center in the early Archaic subsample, is distinctly lower than the center in the Middle Archaic subsample and distinctly higher than the center in the Late Archaic subsample.

In sum, Figure 12.3 shows that, as the centers of the subsamples vary from each other, greater variation is introduced into the overall sample when the three subsamples are combined. Figure 12.2 illustrates a situation where all three subsamples might well have been selected from populations with the same means. Figure 12.3 illustrates a situation where it is considerably more likely that the three subsamples were selected from populations with different means. Analysis of variance finds the key to assessing these probabilities in a comparison between the variance observed *between* subsamples on the one hand and the variance observed *within* subsamples on the other. These two variances, *between groups* and *within groups*, are calculated very much like the variances of ordinary batches of measurements.

Recall the equation for variance from Chapter 3:

$$s^2 = \frac{\Sigma (x - \bar{X})^2}{n - 1}$$

The numerator of this fraction, $\Sigma(x - \bar{X})^2$, is often referred to as the *sum of squares* since it consists of the sum of the squares of the deviations from the sample mean of all the elements in the sample. The denominator, $n - 1$, is actually the number of *degrees of freedom*, a term we did not use in Chapter 3, but which we have come across since.

To calculate the *between groups* variance needed for analysis of variance we must determine what the relevant sum of squares is and what the relevant number of degrees of freedom is. The between groups sum of squares is

$$SS_B = \sum n_i (\bar{X}_i - \bar{X}.)^2$$

where

SS_B = the between groups sum of squares;
n_i = the number of elements in the ith group (or subsample);
$\bar{X}_i$ = the mean in the ith group (or subsample);
$\bar{X}.$ = the mean of all the groups (taken together).

In our example, there are three groups or subsamples, so i refers, in turn, to each of the three groups, whose numerical indexes are given in Table 12.2. Thus

$n_1 = 58$ (the number of Early Archaic projectile points);
$n_2 = 42$ (the number of Middle Archaic projectile points);

n_3 = 27 (the number of Late Archaic projectile points);
$\bar{X}_1$ = 53.67 g (the mean Early Archaic projectile point weight);
$\bar{X}_2$ = 60.45 g (the mean Middle Archaic projectile point weight);
$\bar{X}_3$ = 41.56 g (the mean Late Archaic projectile point weight);
$\bar{X}.$ = 53.34 g (the grand mean projectile point weight, that is, the mean of the total sample including all periods).

Consequently,

$$SS_B = 58(53.67 - 53.34)^2 + 42(60.45 - 53.34)^2 + 27(41.56 - 53.34)^2$$
$$= 6.32 + 2123.19 + 3746.75$$
$$= 5876.26$$

The relevant number of degrees of freedom for this between groups sum of squares is one less than the number of subsamples. For this example, there are three subsamples, so there are two degrees of freedom. Dividing the between groups sum of squares by the number of degrees of freedom, we get

$$s_B^2 = \frac{SS_B}{d.f.} = \frac{5876.26}{2} = 2938.13$$

This figure is the between groups variance, often referred to as the *between groups mean square*. It is the way we express the spread observed between the means of the different groups for analysis of variance.

Analysis of variance seeks to compare the between groups variance just calculated to the within groups variance. This within groups variance, like the between groups variance, involves dividing a sum of squares by the relevant number of degrees of freedom. It amounts to pooling the separate variances of the subsamples. The within groups sum of squares is obtained simply by multiplying each subsample's variance by one less than the number in the subsample and adding up the results for all the subsamples:

$$SS_W = \sum (n_i - 1)s_i^2$$

where

SS_W = the within groups sum of squares;
n_i = the number of elements in the ith group (or subsample) as before;
s_i^2 = the variance of the ith group (or subsample).

Finding the variances of the subsamples in Table 12.2 gives us the following values for s_i^2: $s_1^2 = 215.21$, $s_2^2 = 147.62$, and $s_3^2 = 76.74$. Consequently, in our example,

---

Statpacks

Having gone through the lengthy calculations in the text, we must recognize that calculating an analysis of variance by hand is an outmoded technique. Although there are computational shortcuts that make it easier, and many statistics books provide detailed instructions for these shortcuts, there is not much reason to perform an analysis of variance now except with a computer. This makes it extremely important to understand what an analysis of variance is about, and what the numbers that result from an analysis of variance mean, but it is not important to master the computational details. The point of the example is to help make clear how analysis of variance works, not to provide instructions for how to do it.

Instructions for how to perform analysis of variance can be found in the manual of your computer statpack. Programs do vary, but most will want the data organized as they are in Table 12.1. The period may be called a *grouping variable* or *independent variable* and the weight may be called a *dependent variable*. The output will likely list the between groups and within groups sums of squares, degrees of freedom, and mean squares. The F ratio will be provided, along with its associated probability. Table 12.3 shows an example of one statpack's output for the analysis of this example. This output is from SYSTAT®, and you will note that the numbers are slightly different from those in the text. This is a consequence of rounding error. Statpacks customarily keep track of figures throughout the calculations to many more decimal places than is possible with an ordinary calculator, and thus they produce results with greater precision, although there is no substantive difference in conclusions. Since statpacks also calculate the associated probability with much more accuracy than even a very long and detailed F table can provide, and since it is very unlikely that anyone reading this book will be performing analysis of variance by hand, this book does not include an F table.

$$SS_w = (58 - 1)(215.21) + (42 - 1)(147.62) + (27 - 1)(76.74)$$
$$= 12266.97 + 6052.42 + 1995.24$$
$$= 20314.63$$

The relevant number of degrees of freedom for this within groups sum of squares is the overall sample size minus the number of subsamples. In our example, the overall sample size is 127, and there are three subsamples, so the within groups number of degrees of freedom is 124. Dividing the within groups sum of squares by the number of degrees of freedom, we get

$$s_w^2 = \frac{SS_w}{d.f.} = \frac{20314.63}{124} = 163.83$$

Table 12.3. Example Computer Output for the Analysis of Variance Example in This Chapter

ANALYSIS OF VARIANCE					
SOURCE	SUM OF SQUARES	DF	MEAN SQUARE	F	PROBABILITY
BETWEEN GROUPS	5880.6	2	2940.30	17.94	0.0000001
WITHIN GROUPS	20321.8	124	163.89		

This figure is the within groups variance, often referred to as the *within groups mean square*. It is the expression of the spread to be observed within the various groups needed for analysis of variance.

Once the between groups variance and the within groups variance are calculated, the analysis of variance is almost complete. It only remains to express these two variances as a ratio:

$$F = \frac{s_B^2}{s_W^2}$$

This F ratio in our example comes to

$$F = \frac{2938.13}{163.83} = 17.93$$

The F ratio can then be looked up in a table providing probabilities associated with the different values of F. The probability associated with an F ratio of 17.93 for 2 degrees of freedom between groups and 124 degrees of freedom within groups is 0.0000001. This means that there is only 1 chance in 10 million of randomly selecting three subsamples with the means and standard deviations that these have from three populations whose means are the same. There is, then, a vanishingly small probability that the differences observed between these three samples are simply a consequence of the vagaries of sampling. Our results are extremely significant. We have extremely high confidence that projectile points from different periods really do have different mean weights.

STRENGTH OF DIFFERENCES

In addition to discussing the significance of the differences we observed in the mean weights of projectile points from different parts of the Archaic, we should discuss the strength of these differences. The strength of the differences amounts to nothing more complicated than the differences of means between the subsamples. Late Archaic projectile points are the lightest, on average, with a mean 12.11 g below that of Early Archaic points, which are, in turn, 6.78 g lighter than Middle Archaic ones. Thus the sharpest contrast is the 18.89 g that separate the mean for Middle Archaic points from the mean for Late Archaic points. These are, of course, the differences whose significance we have been evaluating, first by comparing estimated means and error ranges in a bullet graph and later through analysis of variance. Both strength and significance can be seen in Figure 12.1, and that is a clear advantage of such a presentation. It is easy to identify there exactly which subsamples are heavy and which are light and by approximately how much. In fact, most of what we need to say about these numbers is most easily seen in Figure 12.1. For most purposes in archaeology, a bullet graph is much simpler and more straightforward than an analysis of variance. If it is important to put a single probability figure on the entire pattern of subsamples, however, analysis of variance is available.

Whether the results of our analysis are meaningful depends on both significance and strength but in different ways. If there is very little significance, then there is little point in discussing the meaning of the differences observed, because there is too high a probability that we would simply be discussing the random differences between three samples selected from indistinguishable populations. If there is moderate to high significance, then the strength or magnitude of the differences is worth discussing, at least tentatively, because it seems likely that there is a "real" difference to be discussed. If the significance level is very high, then it is worth engaging in serious discussion of the strength of the differences. Even though the significance level is extremely high in our example, this does not automatically make the results meaningful. It makes them very likely to be "real," but many "real" things are trivial. Whether this difference means anything depends on the substantive issues that we are investigating. If smaller projectile points were, indeed, used for hunting smaller animals, then our results might be used to support an interpretation that smaller animals were most hunted in the Late Archaic and larger animals were most hunted in the Middle Archaic, with the Early Archaic falling somewhere in between. This, then, would imply a shift from smaller toward larger and then back toward smaller game. Whether the 10 to 20 g involved in mean weight differences is large enough to be meaningful in this context is a substantive rather than statistical evaluation. And, of course, as always, we would want to look at other completely different lines of evidence relevant to the issue, such as site locations, faunal remains, and many others.

As discussed in Chapter 11, significance and strength are two importantly different concepts. Significance is, in some sense, the more "purely" statistical of the two, while strength usually sets us on the path toward the substantive interpretation of the statistical results. Only when relatively high significance is combined with strong enough results to have substantive meaning do our statistical results have much importance. Highly significant results may have little meaning because they are very weak, and very strong results may have little importance because their significance level is low.

DIFFERENCES BETWEEN POPULATIONS VERSUS RELATIONSHIPS BETWEEN VARIABLES

Analysis of variance can also be thought about from a rather different perspective. Instead of focusing on the differences between several populations in mean values of some measurement, we could focus on the analysis of variance as an investigation of the relationship between two variables. In the example above, the two variables would be projectile point weight and period. In an analysis of variance, conceived in this way, there are always two variables: one of them is a measurement, and the other is a set of categories. It is the categorical variable that provides the basis for the division of the overall sample into subsamples, one corresponding to each category.

The categorical variable is always considered the *independent variable* because we simply take the division of the sample into subsamples based on these categories as a given. The measurement is called the *dependent variable* because we speak as if it were determined, at least in part, by the categories. In the example of Archaic projectile points from the Cottonwood River valley we found that Late Archaic projectile points weighed less, on average, than Early Archaic ones. Thus it seems reasonable to say that projectile point weight *depends on* period to some extent. It is simpler in statistics to speak of the relationship in these terms, although this implies nothing about the direction of causality in the real world. Indeed, it makes little real sense even to talk about period as an independent variable that "causes" projectile points to be larger or smaller. This is simply a convention of statistical language, having little to do with real notions of causality.

It is often useful to think of variable relationships in predictive terms. If the two variables—projectile point weight and period—are related to each other, then knowing the value of one for a particular case would help us to predict the value of the other. If, before looking at a particular projectile point, we wished to predict its weight, the best guess we could make would be the mean of the overall sample. That guess would most often be closest to the real weight of the projectile point in question. Given what we found out in the analysis of variance, however, we know that it would help us make better predictions if we knew to what part of the Archaic the projectile point pertained. If we knew that the point was Late Archaic, the best prediction

---

Be Careful How You Say It

The following sentence provides a complete example of how the conclusions from the example analysis of variance might be stated: "The difference observed in mean weight of Early, Middle, and Late Archaic projectile points in the Cottonwood River valley has extremely high significance ($F = 17.93$, $p = .0000001$)." This tells the reader what you concluded in a meaningful way: it says what significance test was used (because the F ratio is the result of an analysis of variance), and it gives the resulting statistic together with the significance level or associated probability. It would *not* be adequate simply to say, "The difference observed in mean weights of Early, Middle, and Late Archaic projectile points in the Cottonwood River valley is significant." This latter statement is not exactly incorrect, but it is certainly incomplete. It fails to specify what significance test was used, and it gives no information whatsoever about *how* significant the results were. It perpetuates the not very useful idea that being significant (like being pregnant) is a clearcut yes or no condition.

If what we are interested in is more easily framed in terms of the relationship between two variables, then there is yet a different way to phrase the overall conclusion to be drawn from the example analysis of variance: "For Archaic projectile points from the Cottonwood River valley, the relationship between weight and period has extremely high significance ($F = 17.93$, $p = .0000001$)."

One subtlety of reporting significance probabilities from computer output is to recognize what it means if your statpack reports a probability of 0.000. This does not mean absolute certainty. It only means a probability less than 0.0005, since anything greater than or equal to 0.0005 would round off to 0.001 and anything less than 0.0005 would round off to 0.000. Your program may enable you to ask it to show results to more decimal places so that you can see what the probability really is. If not, it is better to say that the probability is less than 0.0005 instead of saying that the probability is 0.000.

---

would be the mean of the Late Archaic subsample. This prediction would more often be closer to the real weight than the prediction based on the overall sample mean. It is in this sense that we can say that knowing the period helps us to predict the projectile point weight. (We could, of course, reverse direction and predict period from weight. It is a little more complicated to phrase, and so we don't usually find it convenient to speak that way, but the relationship is symmetrical in that sense.)

If there were no relationship between projectile point weight and period, then knowing one would not help us predict the other at all. Looked at from this viewpoint, the significance question then becomes, "How likely is it that the relationship between projectile point weight and period that we observe in this sample is simply a consequence of sampling vagaries?" Yet another way to put it would be, "How likely is it that we would select a sample of this size with this strong a relationship between weight and period from a population of projectile points in which the two variables were unrelated?" The analysis of variance answers this question with the F ratio and its associated probability. In our example, the answer to either question is "extremely unlikely," corresponding to only 1 chance in 10 million.

When the question we wish to ask is most naturally framed as one of relationship between two variables rather than differences between populations, then the analysis of variance can provide a convenient single answer to the question. When the question is more naturally framed as one of differences between populations, then the approach by way of estimating means for the different populations and attaching error ranges to the estimates (Figure 12.1) is likely to be much more direct and informative.

ASSUMPTIONS AND ROBUST METHODS

Estimating population means and attaching error ranges to them is strongly affected by outliers. This problem can be corrected by estimating the trimmed mean and attaching an error range to it, as discussed in Chapter 9. It all works exactly the same way, no matter how many subsamples are being compared. Each one is simply treated as an independent sample from which to estimate a population trimmed mean. If the trimmed mean is estimated from one subsample, however, the trimmed mean must be estimated from all subsamples. Comparing a trimmed mean to a regular mean is a comparison of apples and oranges.

Analysis of variance assumes that the samples are drawn from populations with normal shapes and that the spreads of the populations are approximately equal. Means of checking the validity of these assumptions, relying largely on stem-and-leaf plots and box-and-dot plots, were discussed at the beginning of the example analysis above. These assumptions will be recognized as precise parallels to the assumptions of the two-sample t test. If the spreads in the subsamples are very different, then that is, in itself, an indication that they did not come from identical populations. If the shapes of the subsamples are very asymmetrical, then a transformation that produces reasonably symmetrical shapes for all subsamples can be applied.

If the subsamples to be compared contain outliers, the analysis of variance can be based on the trimmed means and trimmed standard deviations, as discussed in Chapters 2 and 3. Few computer programs provide this as an option in analysis of variance, but it is not difficult to use most statistics packages to help you arrive at the trimmed mean and trimmed standard deviation for each subsample. Once these figures have been obtained, you have information analogous to that in Table 12.2, which you can use to calculate the final steps in the analysis of variance by hand as discussed in the text, simply using the trimmed mean and the trimmed standard deviation squared wherever the regular mean and the regular standard deviation squared are called for. (You will, of course, then need to go find an F table to look up the probability associated with the statistic you produce. Many statistics books contain this table.)

PRACTICE

You are interested in investigating variability in group mobility, which you think is related to the size of the house that a family builds. You have excavated a series of Neolithic houses at five different sites near Heiligenstadt. Each site is in a different environmental setting, but each was occupied through all three parts of the Neolithic that you can identify: Early, Middle, and Late. The information is given in Table 12.4. A long Oktoberfest recess in your field season provides ample opportunity for deep consideration of issues of sampling bias, and you decide that you will use the sample of house floors from each site as a random sample from a much larger and vaguely defined population consisting of all house floors from environmental settings like that of the site in question. Likewise, you will use the sample of house floors from Early, Middle and Late Neolithic as a random sample from the large and vaguely defined population of all house floors of that period.

1. Estimate the mean house floor area in each of the five environmental settings represented by the five different sites. Draw a bullet graph comparing these five populations in regard to estimated mean house floor area with error ranges for 80%, 95%, and 99% confidence levels. Does it appear that house sizes were different in different environmental settings? Summarize what you can conclude from your graph in one or two clear sentences.

2. Perform an analysis of variance to evaluate the relationship between house floor area and site based on this sample of 76 house floors. Does it appear that there is a relationship between environmental setting and house size? State the results of your analysis in one clearly worded sentence.

3. Estimate the mean house floor area for the region in each part of the Neolithic period. Draw a bullet graph comparing the Early, Middle, and Late Neolithic in regard to house size with error

Table 12.4. Data on House Floor Area for Five Sites Occupied during the Early, Middle, and Late Neolithic near Heiligenstadt

Floor area (m^2)	Site	Neolithic subperiod	Floor area (m^2)	Site	Neolithic subperiod
19.00	Hlg001	Early	15.94	Hlg002	Middle
16.50	Hlg004	Middle	23.05	Hlg003	Middle
16.10	Hlg002	Late	24.15	Hlg001	Early
19.20	Hlg001	Late	20.35	Hlg003	Middle
15.20	Hlg005	Middle	18.95	Hlg004	Early
20.40	Hlg001	Middle	16.85	Hlg002	Middle
16.40	Hlg002	Early	19.95	Hlg003	Early
16.40	Hlg002	Late	20.16	Hlg001	Early
16.40	Hlg002	Middle	19.16	Hlg003	Middle
15.40	Hlg005	Early	17.66	Hlg004	Early
20.60	Hlg001	Middle	15.26	Hlg001	Middle
17.20	Hlg004	Middle	16.26	Hlg005	Middle
19.90	Hlg003	Late	19.46	Hlg002	Late
22.01	Hlg001	Late	15.46	Hlg005	Early
21.11	Hlg003	Early	18.66	Hlg002	Early
16.51	Hlg002	Early	18.36	Hlg004	Middle
22.71	Hlg003	Middle	16.07	Hlg005	Late
20.81	Hlg001	Late	17.17	Hlg002	Middle
15.81	Hlg005	Late	17.17	Hlg003	Late
16.52	Hlg004	Early	20.47	Hlg003	Late
21.12	Hlg003	Late	23.57	Hlg003	Late
18.22	Hlg001	Late	22.77	Hlg001	Middle
23.22	Hlg003	Early	22.77	Hlg003	Late
16.32	Hlg005	Late	15.87	Hlg005	Late
16.13	Hlg005	Early	15.08	Hlg005	Middle
15.33	Hlg002	Early	18.28	Hlg001	Early
16.83	Hlg004	Early	15.78	Hlg004	Late
16.43	Hlg004	Middle	16.98	Hlg004	Late
13.04	Hlg002	Late	20.58	Hlg001	Early
21.14	Hlg003	Middle	16.08	Hlg004	Early
18.24	Hlg005	Early	21.68	Hlg003	Middle
17.34	Hlg002	Late	15.09	Hlg005	Late
14.84	Hlg004	Middle	17.79	Hlg004	Late
17.34	Hlg005	Middle	17.09	Hlg004	Late
21.64	Hlg001	Early	21.69	Hlg001	Middle
15.74	Hlg005	Early	21.69	Hlg001	Late
19.84	Hlg001	Middle	20.69	Hlg003	Early
22.99	Hlg003	Early	24.99	Hlg003	Early

ranges for 80%, 95%, and 99% confidence levels. Does it appear that house size changed through time? Summarize what you can conclude from your graph in one or two clear sentences.

4. Perform an analysis of variance to evaluate the relationship between house size and period based on this sample of 76 house

floors. Does it appear that there is a relationship between house size and period? State the results of your analysis in one clearly worded sentence.

Chapter 13

Comparing Proportions of Different Samples

Sometimes we have a sample divided into subsamples as in the example in Chapter 12, but the comparison we wish to make between the subsamples concerns not the mean of some measurement but rather another set of categories. Such a comparison can be approached by estimating population proportions from the various subsamples and attaching error ranges to the estimates. Then the estimated population proportions with their error ranges can be compared to each other with a bullet graph just as we did for means in Chapter 12.

COMPARISON WITH ESTIMATED PROPORTIONS AND ERROR RANGES

Table 13.1 provides some information about the quantities of sherds of two different vessel forms (bowls and jars) found at two sites (San Pablo and San Pedro). After carefully considering issues of sampling bias we decide that the methods by which these surface collections were made allow us to treat

Table 13.1. Sherds of Different Vessel Forms from the San Pablo and San Pedro Sites

	Bowl sherds	Jar sherds	Total
San Pablo	18	12	30
San Pedro	18	22	40
Total	36	34	70

them as if they were random samples from the large and vaguely defined populations consisting of all the sherds at each site. We calculate the proportions of bowl and jar sherds in each sample and use these proportions as estimates of the corresponding population proportions, attaching error ranges to them on the basis of their standard errors, as discussed in Chapter 10. The estimate for the San Pablo site is 60% bowl sherds and 40% jar sherds, with a standard error of 9% for both. The estimate for the San Pedro site is 45% bowl sherds and 55% jar sherds, with a standard error of 8% for both.

These results are illustrated with a bullet graph in Figure 13.1. Only the proportions of bowl sherds are graphed, since the bullet graph for jar sherds would show exactly the same contrast between the two sites in reverse. Based on these samples, we would say that the San Pablo site has a higher proportion of bowl sherds than the San Pedro site. Our confidence in this statement, however, would not be very high. Comparing the error ranges for different levels of confidence reveals that the estimated proportion for the San Pedro site falls well within the 99% confidence error range for the San Pablo site, and vice versa. Thus our confidence that our samples actually reflect a difference between the two sites (as opposed to reflecting just the vagaries of sampling) is less than 99%. Continuing the comparison, we note that the estimated propor-

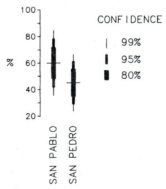

Figure 13.1. Comparison of bowl sherd proportions at the San Pablo and San Pedro sites.

tion for the San Pedro site also falls within the 95% confidence error range for the San Pablo site and vice versa. Thus our confidence that our samples actually reflect a difference between the two sites is even less than 95%. The proportion for the San Pedro site does, however, fall outside the 80% confidence error range for the San Pablo site and vice versa. Thus our confidence that the observed difference reflects something more than just sampling vagaries is somewhere between 80% and 95%—moderate but not very high confidence. The difference might well be strong enough to be meaningful (a difference between 45% and 60%), but the risk that the difference might reflect nothing more than the chance variation between two relatively small samples from identical populations is higher than we might like.

COMPARISON WITH CHI-SQUARE

We first approached the comparison of a measurement between two or more samples by estimating population means (Chapters 11 and 12) and then turned to significance tests that boiled the entire comparison down to a single probability value (the t test and analysis of variance). We have now, in similar fashion, approached the comparison of a set of categories between two samples by estimating population proportions. In this instance, too, there is a significance test that sums the entire comparison up in a single probability value. It is the chi-square test, named after the statistic that it produces, χ^2, represented by the Greek letter chi. The chi-square test works for any number of categories into which the overall sample is divided and for any number of categories for which proportions are calculated. Thus, for proportions, unlike means, there is no division between the two-sample case, where we used the t test to compare measurements, and the multiple-sample case, where we used analysis of variance for the same purpose.

Table 13.1 is easily recognized as the kind of table we worked with in Chapter 6. It seems natural to look at this table in terms of row proportions, because the rows are the two sites and it is the two sites that we want to compare to each other to investigate whether, for example, a difference in activities between the two sites might be reflected in different proportions of ceramic vessel forms. This is what we have, in fact, already been doing in comparing bowl proportions between the two sites. Table 13.2 provides these row proportions. We can see that the San Pablo site has a higher-than-average proportion of bowls while the San Pedro site has a lower-than-average proportion of bowls—just what we concluded from Figure 13.1. We could represent these departures from average with bar graphs as we did in Chapter 6, but this is such a simple comparison that it hardly seems necessary.

Chi-square is based on an assessment of these departures from average. This is accomplished by constructing a table of *expected* values to compare

**Table 13.2. Row Proportions of Sherds
of Different Vessel Forms from the
San Pablo and San Pedro Sites**

	Bowl sherds	Jar sherds	Total
San Pablo	60.0%	40.0%	100.0%
San Pedro	45.0%	55.0%	100.0%
Average	51.4%	48.6%	100.0%

with the table of *observed* values (Table 13.1). If the average proportion of bowl sherds is 51.4%, as indicated in Table 13.2, then we would, in some sense, expect both the San Pablo and San Pedro sites to have 51.4% bowls. For the San Pablo site, this means 51.4% of 30 sherds, or 15.42 bowl sherds. For the San Pedro site, 51.4% of 40 sherds is 20.56 bowl sherds. Correspondingly, we would expect both sites to have 48.6% jar sherds. These expected values are shown in Table 13.3.

Notice that the row and column totals (known together as the *marginal totals*) stay the same (allowing for rounding error) in the table of expected values as they were in the table of observed values. Indeed, it is the constant marginal totals upon which the expected values are based. The short cut for computing the expected values, in fact, is to multiply the row total corresponding to a particular cell by the column total corresponding to that cell and divide by the grand total for the table. For example, to obtain the expected number of bowl sherds at the San Pablo site, we could multiply the row total for that cell (30) by the column total for that cell (36) and divide by the grand total (70) to obtain 15.43—exactly what we obtained from the row proportions (allowing for rounding error). We arrive at the same expected values for the table no matter whether we use row proportions, column proportions, or multiplication of marginal totals. This table of expected values provides the basis for a summary statistic, χ^2.

The χ^2 statistic is really very like a standard deviation, in that it involves calculating deviations, squaring them, and summing them up. The

**Table 13.3. Expected Number of Sherds
of Different Vessel Forms from the
San Pablo and San Pedro Sites**

	Bowl sherds	Jar sherds	Total
San Pablo	15.42	14.58	30
San Pedro	20.56	19.44	40
Total	36	34	70

deviations, however, instead of being deviations from the mean, as they are for the standard deviation, are observed deviations from expected values:

$$\chi^2 = \sum \frac{(O_i - E_i)^2}{E_i}$$

where

O_i = the observed value for the ith cell of the table;
E_i = the expected value for the ith cell of the table.

Our example is what is often referred to as a *two-by-two table* because it has two rows and two columns. There are, therefore, four cells. We thus calculate the quantity $(O_i - E_i)^2/E_i$ for each of the four cells and sum up the four values:

$$\chi^2 = \frac{(18 - 15.42)^2}{15.42} + \frac{(12 - 14.58)^2}{14.58} + \frac{(18 - 20.56)^2}{20.56} + \frac{(22 - 19.44)^2}{19.44}$$
$$= 0.4317 + 0.4565 + 0.3188 + 0.3371$$
$$= 1.5441$$

This value, $\chi^2 = 1.5441$, is then looked up in Table 13.4 to determine the associated probability. One need only determine the appropriate number of degrees of freedom, which for χ^2 is the product of one less than the number of rows in the table times one less than the number of columns in the table. Since the table in our example has two rows and two columns, the number of degrees of freedom is $1 \times 1 = 1$. Using the first row in the table, then, for one degree of freedom, we see that the χ^2 value of 1.544 falls between the table values of .455 and 1.642. Thus the associated probability is between 50% and 20%. As with other significance tests, this is the probability that the differences we observe (in this case between the two sites in regard to proportions of sherds of different vessel forms) are a consequence of the vagaries of sampling—that is to say, the probability that we could select two samples with proportions as different as these from parent populations having identical proportions. The chi-square test, then, is designed to answer the question, "How likely is it that we could select samples with proportions of bowl and jar sherds as different as these if the two sites did not really differ in regard to bowl and jar sherd proportions?"

In this example, our answer to the question is that there is somewhere between a 50% and a 20% risk that we could select samples as different as these if the two sites did not really differ in regard to bowl and jar sherd proportions. This is a high enough risk that our samples do not "really" indicate any difference between the two sites that we would not regard this evidence as

Table 13.4. The Chi-Square Distribution[a]

Confidence	50%	80%	90%	95%	98%	99%	99.9%
	.5	.8	.9	.95	.98	.99	.999
Significance	50%	20%	10%	5%	2%	1%	0.1%
	.5	.2	.1	.05	.02	.01	.001
Degrees of freedom							
1	.455	1.642	2.706	3.841	5.412	6.635	10.827
2	1.386	3.219	4.605	5.991	7.824	9.210	13.815
3	2.366	4.642	6.251	7.815	9.837	11.341	16.268
4	3.357	5.989	7.779	9.488	11.668	13.277	18.465
5	4.351	7.289	9.236	11.070	13.388	15.086	20.517
6	5.348	8.558	10.645	12.592	15.033	16.812	22.457
7	6.346	9.803	12.017	14.067	16.622	18.475	24.322
8	7.344	11.030	13.362	15.507	18.168	20.090	26.125
9	8.343	12.242	14.684	16.919	19.679	21.666	27.877
10	9.342	13.442	15.987	18.307	21.161	23.209	29.588
11	10.341	14.631	17.275	19.675	22.618	24.725	31.264
12	11.340	15.812	18.549	21.026	24.054	26.217	32.909
13	12.340	16.985	19.812	22.362	25.472	27.688	34.528
14	13.339	18.151	21.064	23.685	26.873	29.141	36.123
15	14.339	19.311	22.307	24.996	28.259	30.578	37.697
16	15.338	20.465	23.542	26.296	29.633	32.000	39.252
17	16.338	21.615	24.769	27.587	30.995	33.409	40.790
18	17.338	22.760	25.989	28.869	32.346	34.805	42.312
19	18.338	23.900	27.204	30.144	33.687	36.191	43.820
20	19.337	25.038	28.412	31.410	35.020	37.566	45.315
21	20.337	26.171	29.615	32.671	36.343	38.932	46.797
22	21.337	27.301	30.813	33.924	37.659	40.289	48.268
23	22.337	28.429	32.007	35.172	38.968	41.638	49.728
24	23.337	29.553	33.196	36.415	40.270	42.980	51.179
25	24.337	30.675	34.382	37.652	41.566	44.314	52.620
26	25.336	31.795	35.563	38.885	42.856	45.642	54.052
27	26.336	32.912	36.741	40.113	44.140	46.963	55.476
28	27.336	34.027	37.916	41.337	45.419	48.278	56.893
29	28.336	35.139	39.087	42.557	46.693	49.588	58.302
30	29.336	36.250	40.256	43.773	47.962	50.892	59.703

[a]Adapted from Table 14 in *Tables for Statisticians* by Herbert Arkin and Raymond R. Colton (New York: Barnes and Noble, 1963).

much support for the notion of a difference in activities between the two sites. This is a slightly different conclusion than we came to from the bullet graph in Figure 13.1. By looking at the bullet graph, we decided we would have between 80% and 95% confidence that there was a difference in bowl and jar sherd proportions between the two sites. A confidence level between 80% and 95% ought to translate into a significance probability between 20% and 5%, but the chi-square test gave us a significance probability between 50% and 20%. This is because the two approaches are not just mirror image applications

Degrees of Freedom

In the tables on which the chi-square statistic is based, the term *degrees of freedom* makes some intuitive sense. There are, of course, numerous ways to fill cell values into a table so that they add up to a given set of marginal totals. In a two-by-two table, however, once a single cell value has been filled in, the other three cell values are determined because there is only one value for each of the other three cells that will make the given marginal totals add up correctly. This is, in a sense, the one degree of freedom that a two-by-two table has. For, say, a three-by-four table, there are six degrees of freedom (one less than the number of rows times one less than the number of columns), and it takes six cell values to completely determine such a table for a given set of marginal totals. (Try this out on paper, and you'll soon see just how it works. There is no set of five cells or fewer in a three-by-four table whose values will completely determine what the rest of the cell values must be to produce a given set of marginal totals. It takes six.)

Thinking back to calculations of standard deviations in sample batches of numbers and to the use of the *t* table reveals a related principle. For the *t* table, the number of degrees of freedom is one less than the number in the sample. If a sample batch has a given mean, then it is necessary to establish what all the numbers but one are before the last number is constrained to a single value. There is, of course, much more mathematical logic to this notion, but degrees of freedom in using the *t* table, in using the chi-square table, and, for that matter, in dividing by $(n - 1)$ in calculating the standard deviation of a sample are related to this notion.

of the same principles, as the bullet graph and the *t* test in Chapter 11 were. Here the error ranges in the bullet graph and the chi-square test are taking slightly different approaches, and it is not surprising that they produce slightly different results—yes, slightly different, because the two results are not really as different as they seem. If we look carefully at the bullet graph, we can see that the confidence we should have really is closer to 80% than to 95%. And if we look carefully at the chi-square table, we see that our result really is much closer to the 20% significance column than the 50% significance column. Thus, the bullet graph suggests slightly greater than 80% confidence while the chi-square test suggests a significance slightly greater than 20%, so the two results do not in fact disagree by very much.

MEASURES OF STRENGTH

Just as in the other situations we have discussed, the significance and the strength of a result are different things. In this example, it is the signifi-

Be Careful How You Say It

In conclusion to the chi-square example in the text, we can say, "The difference between the San Pablo site and the San Pedro site with respect to proportions of bowl sherds and jar sherds is not very significant ($\chi^2 = 1.544$, $.50 > p > .20$)." This statement makes clear just what differences were investigated; it informs the reader what significance test was used, since χ^2 is the result of the chi-square test; and it provides the reader with the resulting statistic and its associated probability.

It would *not* be adequate to conclude this significance test simply by saying, "The San Pablo site and the San Pedro site do not differ significantly in proportions of bowl and jar sherds." In the first place, this latter statement does not tell the reader what significance test was used or provide its specific results. In the second place, it treats significance as a simple yes or no condition, which is at the least an oversimplification. On this last score, the inadequate statement even tends to mislead. The χ^2 value obtained (1.544) actually falls fairly close to the 20% significance column. Interpolating from the table, then, the actual probability must be only slightly greater than 20%. Put another way, the confidence we have that the two sites actually differ in bowl and jar sherd proportions is somewhere near 80%. We should, then, be saying that there is almost an 80% chance that the differences observed between the two samples actually do reflect differences between the sites rather than just the vagaries of sampling. The risk is still substantial (slightly over 20%) that nothing more may be at work here than the random variation of samples, but it is certainly more likely that there actually are differences in bowl and jar sherd proportions between the two sites. The last thing we want to do on the basis of this significance test is to act as if we have established that the two sites have the same proportions of bowl and jar sherds. This is why it is worth communicating that the odds favor the conclusion that there *is* a difference between the sites, even though there remains a worrisomely large risk that this may not be the case.

Under the influence of the near-sacred 5% significance level for rejecting or failing to reject the null hypothesis (see Chapter 11), people are accustomed to characterizing significance levels around 5% as "high." When the significance level reaches 1% or less, it is common to characterize it as "very high." A significance level around 20%, as in the example in the text, would usually be called "low." While it may seem that we've gone all the way from "very high" to "low" while staying pretty much toward one end of the scale, it is reasonable to think in such terms because, once the significance level goes far above 20%, the risk that the samples differ only because of the vagaries of sampling is so great that the result merits little attention. A difference between samples that is "highly significant" corresponds to "high confidence" that there is a difference. Note, though, that "high" significance corresponds to low associated significance probabilities (say, 5% or less), and "low" significance corresponds to high associated significance probabilities (say, around 20% or greater).

cance level that gives us serious pause. There is somewhere around a 20% probability that we could select random samples and get the results that we got even if the two sites in fact had identical proportions of bowl and jar sherds. This is certainly far from reassuring. On the other hand, it is more likely that the difference observed between the two samples actually reflects a difference between the two sites. If this is the case, then the difference observed (15%) is probably strong enough to have a meaningful interpretation. Unlike the t test and analysis of variance, several specific measures of strength of results come along with the χ^2 test's measure of significance.

One of the most flexible and easiest to calculate is Cramer's V:

$$V = \sqrt{\frac{\chi^2}{n\,(S-1)}}$$

where

n is the number of elements in the sample (that is, the grand total for the table);
S is the number of rows or the number of columns in the table, whichever is smaller.

Thus, in our example,

$$V = \sqrt{\frac{1.544}{70(1)}} = 0.15$$

It can be shown that V ranges from zero to one. It takes on a value of zero when there is no difference at all between the observed values and the expected values, and it takes on a value of one when the difference between observed and expected values is as large as it can be. This latter would occur, for example, if the San Pedro site had only bowl sherds and the San Pablo site had only jar sherds. Thus, the closer V is to one, the stronger is the difference in proportions between the categories. (For a two-by-two table, V is the same as the difference between the observed proportions—here a difference of 15% between 60% and 45% for bowl sherds or 55% and 40% for jar sherds.)

For any table that has no more than two rows (or alternatively, no more than two columns), the value of $S-1$ will always be 1, and this term will have no effect on the outcome. In this situation, V is equivalent to another measure of strength, called ϕ (the Greek letter phi, which in statistics is usually pronounced "fee"). The calculation of ϕ is quite simple: divide the χ^2 score by the grand total of the table and take the square root of the result. As long as the table is only two rows or only two columns, ϕ is limited to a range between zero and one and is exactly the same thing as V. For tables with more than two rows or more than two columns, ϕ is not very useful because its range becomes open-ended. V can be thought of as a modification of ϕ, expanding its utility to tables of any size. It is convenient simply to use V for tables of any size and to recall that when someone refers to ϕ for a table of two rows or two columns it is the same thing as V.

THE EFFECT OF SAMPLE SIZE

Having obtained the results we obtained in the example χ^2 test, we might decide that the possibility of differences between the two sites is intriguing, and we might want to explore it further. The very modest significance of the results, of course, was in part attributable to the fact that our samples were relatively small. (It is more likely that small samples will differ widely from the populations they are selected from than that large samples will. Thus the likelihood of getting a large difference between two small samples purely because of the vagaries of sampling is greater.) We might thus decide to seek larger samples of sherds from the two sites. Table 13.5 provides some imaginary results of seeking larger samples. Now there are exactly four times as many sherds, in exactly the same proportions as before. The strength of differences in proportions, then, remains the same (15%). Table 13.2 provides row proportions that are equally valid for the new result. The new expected values (Table 13.6) are, likewise, four times the old expected values (Table 13.3).

Calculating the χ^2 score, though, on the basis of the larger sample gives a very different result:

$$
\begin{aligned}
\chi^2 &= \frac{(72-61.71)^2}{61.71} + \frac{(48-58.29)^2}{58.29} + \frac{(72-82.29)^2}{82.29} + \frac{(88-77.71)^2}{77.71} \\
&= 1.7158 + 1.8165 + 1.2867 + 1.3626 \\
&= 6.1816
\end{aligned}
$$

Table 13.5. A Larger Sample of Sherds of Different Vessel Forms from the San Pablo and San Pedro Sites

	Bowl sherds	Jar sherds	Total
San Pablo	72	48	120
San Pedro	72	88	160
Total	144	136	280

**Table 13.6. Expected Numbers of Sherds
of Different Vessel Forms from the San Pablo
and San Pedro Sites with a Larger Sample**

	Bowl sherds	Jar sherds	Total
San Pablo	61.71	58.29	120
San Pedro	82.29	77.71	160
Total	144	136	280

A χ^2 score of 6.1816 for one degree of freedom is associated with a significance level between .02 and .01. These results are highly significant. We might say, based on this sample, that we are between 98% and 99% confident that there are differences of bowl and jar sherd proportions between the San Pablo and San Pedro sites. The very different character of this conclusion from the one we reached before is attributable simply to sample size. Other things being equal, results from larger samples are more significant than results from smaller samples. The strength of the difference in proportions is still the same (a 15% difference between the sites in the proportion of bowl or jar sherds). And V continues to be 0.15.

For very small samples, only very strong results turn out to be significant. For larger samples even weaker results can be more significant. And for very large samples, even very weak results can have extremely high significance. Strength of results is most closely connected to meaning. It seems likely that the 15% difference in bowl sherd proportion between these two sites reflects some difference in the use of ceramic vessels at the two sites. Would a difference of only 5% have a similar meaning? Of 1%? Of 0.1%? At some point we would surely say that the difference in proportions was so weak that it meant little in terms of differences in ceramic vessel use. And yet we could certainly acquire a sample large enough to find even a tiny difference of 0.1% highly significant. It would clearly, however, not be worth the effort of acquiring such a large sample because it would not (at least in this regard) tell us anything useful. *Large samples are not necessarily more informative than smaller samples, because they may simply increase the statistical significance of results that are too weak to be meaningful.*

Precisely the same contrast between smaller and larger samples can, of course, be seen in estimates of population proportions like those illustrated in Figure 13.1. The effect of the larger sample on that bullet graph would be to shorten the error bars substantially so that the difference in bowl proportions (which would remain the same) would be considerably more significant. If this does not seem intuitively sensible to you, try it out for yourself. Make a revised bullet graph with error bars calculated from the larger sample, and you will see just exactly how increasing the number of elements in the sample narrows in the error ranges for any given level of confidence.

DIFFERENCES BETWEEN POPULATIONS VERSUS RELATIONSHIPS BETWEEN VARIABLES

Just as analysis of variance can be thought of either as a study of differences between populations or as an investigation of relationship between variables, so can a χ^2 test. In this instance, both variables are categorical. We would call the ones in the example analysis something like "site," with two categories (San Pablo and San Pedro), and "vessel form," also with two categories (bowls and jars). We have found an *association* between these two variables—an association of some strength but little significance. If we had framed the analysis as one investigating the relation between two variables rather than the difference between two populations, then we might conclude, "Vessel form proportions do differ somewhat from one site to the other, but there is little significance to the relationship between site and vessel form ($\chi^2 = 1.544$, $.50 > p > .20$, $V = 0.15$)."

As in Chapters 11 and 12, the bullet graph is an alternative to a significance test. For many purposes, the bullet graph in Figure 13.1 would be the clearest and most straightforward way to present the observed differences between the San Pedro and San Pablo samples and the confidence we have in those observations. It would be statistical overkill to present such a bullet graph and then go on to present the results of a chi-square test. (It would indeed be a waste of time to calculate both.) They tell the same story. Pick the version that most serves the need at hand and use it.

Chi-square tests usually get more and more difficult to interpret meaningfully as the number of categories (and thus the number of rows and columns in the table) increases. With so many cells, it is usually only a few that show big differences between observed and expected values, and various techniques have been suggested for homing in on which specific cells in a large table really have important differences. They all boil down to comparing observed and expected values in some way, and for this reason it is common to include tables of observed and expected values to accompany chi-square results. It is often preferable to use bullet graphs in such a case, since they portray the categories individually, with the separate confidence/significance implications of each included in the graph.

ASSUMPTIONS AND ROBUST METHODS

In relating one categorical variable to another categorical variable, there are none of the problems inherent in the use of means and standard deviations on measurement variables. Thus assumptions concerning shapes are not made, and worries about the effect of outliers simply do not arise. The

Statpacks

When framing a χ^2 analysis as a test of relationship between two variables and when the data on the two categorical variables are organized as they are in Table 6.1, computer statpacks are likely to be of considerable assistance in performing χ^2 tests. If the data are already in the form of a table like that in Table 6.4 or Table 13.1, then calculation of the χ^2 score is probably more easily accomplished by hand. Most of the work is in counting up the numbers for the table, and it is under the heading of *cross tabulations* that many statpacks deal with chi-square. Just by way of comparison with the example we calculated in this chapter, a χ^2 test performed on the data from Table 6.1 with a statpack indicates that the differences in proportions of incised and unincised ceramics from site to site are of moderate strength and little significance ($\chi^2 = 2.493$, $p = .29$, $V = 0.133$). The example from Table 13.1 in this chapter yields the same results when performed with a statpack that we already calculated by hand, except that, as usual with a statpack, the associated probability is calculated more precisely, $p = .214$, confirming the rough interpolation we made from Table 13.4 that the significance probability was between 20% and 50% but much closer to 20%.

principal concern about the χ^2 test is that the sample be large enough for it to be reliable. Many different rules of thumb can be found concerning this in different statistics texts. Some statisticians would like us not to use the χ^2 test if any of the expected values in the table are less than 10. Others are much less conservative and are willing to accept χ^2 tests based on tables with expected values as low as 1. A middle course, one that we will adopt here, is to insist that no expected value be less than 1 *and* that no more than 20% of the expected values be less than 5.

If these conditions are not met, and the table is a large one (that is, a table with many rows and/or many columns), it is often feasible to combine categories for one or both variables, so that there are fewer rows and/or fewer columns in the table. With the same number of cases divided among fewer cells, the expected values, of course, will be higher. Since we have adopted a relatively unconservative requirement for expected values, combining categories will ordinarily suffice to bring these expected values up to acceptable levels.

If a two-by-two table has such low expected values that the results of the chi-square test are unreliable, there is an alternative in the form of Fisher's exact test. This test is a direct calculation of the significance probability and there are no requirements at all about how large the expected values need be. Indeed, expected values do not even enter into its calculation. Fisher's method for calculating the exact significance probability for a two-by-two table is

$$p = \frac{(A + B)! \; (C + D)! \; (A + C)! \; (B + D)!}{N! \, A! \, B! \, C! \, D!}$$

where

A = the observed frequency in the upper left cell of the two-by-two table;
B = the observed frequency in the upper right cell of the two-by-two table;
C = the observed frequency in the lower left cell of the two-by-two table;
D = the observed frequency in the lower right cell of the two-by-two table.

The mathematical use of the symbol ! may not be familiar. X! (read "X factorial") means to multiply X sequentially by each positive integer less than X. For example $5! = 5 \times 4 \times 3 \times 2 \times 1 = 120$. Or $9! = 9 \times 8 \times 7 \times 6 \times 5 \times 4 \times 3 \times 2 \times 1 = 362,880$. Calculating the probability for the example in Table 13.1 yields

$$p = \frac{(18 + 12)! \; (18 + 22)! \; (18 + 18)! \; (12 + 22)!}{70! \, 18! \, 12! \, 18! \, 22!} = 0.237$$

This is a calculation that most people will be willing to leave to their computers, but it does provide the exact significance probability for the example from Table 13.1—a probability for which the χ^2 test gives only an approximation. Most important, Fisher's exact test can be applied regardless of how low the expected cell values are, and when the numbers are small, the calculations are less formidable for those doing them by hand.

POSTSCRIPT: COMPARING PROPORTIONS TO A THEORETICAL EXPECTATION

Sometimes one arrives at a question in data analysis quite similar to the question we have been dealing with up to now in this chapter, but with one important difference. Perhaps our sample can be divided up into a set of categories and we know what we expect the proportions in those categories to be—based not on the proportions in another set of categories into which our sample can be divided, but rather on some different criterion. For example, suppose that we have results of regional survey in three different environmental settings as given in Table 13.7. Since most of the territory surveyed was in the river bottoms, we might expect to find most of the sites in that setting, other things being equal. As Table 13.7 makes clear, however,

Table 13.7. Regional Survey in Three Environmental Settings

Environmental setting	No. of Sites	% of total sites	Area surveyed (km^2)	
			km^2	% of total
Remnant levees	19	50.0	3.9	28.7
River bottoms	12	31.6	8.3	61.0
Slopes	7	18.4	1.4	10.3
Total	38	100.0	13.6	100.0

the proportions of sites in the three settings are quite different from what we might expect. But is our sample large enough to give us much confidence in these differences from our expectations?

At first glance, one might be tempted to answer this question with a chi-square test beginning as in Table 13.8. Thinking about that table, however, should make us pause. Table 13.8 does not really involve two variables that are two separate sets of categories for dividing up the same sample in two different ways. Instead it only involves one set of categories (the three environmental settings). Two different things have been divided up according to this one set of categories—the 38 sites and the 13.6 km^2 of surveyed area. It makes no sense at all to add up 38 and 13.6 and say that we have a sample of 51.6. (51.6 what?) Yet that is what we would be doing if we simply used the numbers in Table 13.8 to calculate χ^2.

In this example, what we have is a sample of 38 sites. If we are willing to treat this as a random sample of sites in the region, then we can compare the proportions of sites in different settings to the theoretical expectation based on how the territory surveyed to find the 38 sites was divided between the three different settings. One way to do this would be to follow the approach discussed in Chapter 10 for estimating the proportions of sites in different settings and for attaching error ranges to these estimates. This would tell us, for instance, that we can have 99% confidence that our sample of 38 sites came from a population in which 31.6% ± 20.3% of the sites were located in the river bottoms. (You can calculate this yourself, following the procedure in Chapter 10.) If prehistoric inhabitants showed no preference for

Table 13.8. An *Incorrect* Way to Tabulate Observed Values from Table 13.7 for a Chi-Square Test

	No. of sites	Area surveyed (km^2)
Remnant levees	19	3.9
River bottoms	12	8.3
Slopes	7	1.4

any of these environmental settings, however, the proportion we would have expected here was 61.0%, since 61.0% of the territory is in this zone. This proportion is considerably higher than the top of the 99% confidence error range (51.9%), and so we can say that it is extremely unlikely that our sample came from a population with 61.0% sites in the river bottoms. Although we do find sites there, it seems that the prehistoric inhabitants showed something of an aversion to settling in the river bottoms. (Or possibly that recent sedimentation has covered more sites in the river bottoms than elsewhere, resulting in a particular failure to discover such sites on survey. This is a question of interpretation that statistical analysis of these numbers will not help us with.)

If we want to know exactly how unlikely it is that our sample came from a population with 61.0% sites in the river bottoms, we could perform a one-sample t test, as discussed in Chapter 11. This example is different from the example one-sample t test in Chapter 11 only in that there are three categories involved rather than just two. We could follow this approach to each of the three categories, determining whether the proportion estimated on the basis of our sample was greater or less than we would expect and how likely it was that the difference observed could be attributed to the amount of random variation ordinarily seen in samples the size of ours. This would lead to a specific discussion of settlement preferences (or apparent lack thereof) in regard to each of the three environmental settings.

We might want to treat the issue in a more comprehensive way, however, focusing not on each individual category, but asking the more general question, "How likely is it that this entire sample of 38 sites came from a population of sites in which there was no preference for locating sites in any particular environmental setting?" We can use a chi-square test to answer this question, but not in the way indicated in Table 13.8. Instead, we use the information we have to determine expected numbers of sites in each environmental setting as in Table 13.9. Since 28.7% of the surveyed area was on remnant levees, we might expect 28.7% of the 38 sites found (10.9 sites) to be on remnant levees, and so on. We now have a one-variable tabulation—observed and expected values for three categories. We can use these observed and expected values to calculate χ^2 just as before:

$$\chi^2 = \sum \frac{(O_i - E_i)^2}{E_i}$$

$$\chi^2 = \frac{(19 - 10.9)^2}{10.9} + \frac{(12 - 23.2)^2}{23.2} + \frac{(7 - 3.9)^2}{3.9}$$

$$= 6.0192 + 5.4069 + 2.4641$$

$$= 13.8902$$

Table 13.9. Observed and Expected Numbers of Sites for Chi-Square Test

		No. of Sites	
	Area surveyed	Exp.	Obs.
Remnant levees	28.7%	10.9	19
River bottoms	61.0%	23.2	12
Slopes	10.3%	3.9	7
Total	100.0%	38	38

Since there is only one row in this table (or one column—it makes no difference whether the table is vertical or horizontal), the number of degrees of freedom is one less than the number of categories. Here, there are three categories, so two degrees of freedom. A value of 13.8902 for χ^2 is just beyond the rightmost column of Table 13.4 in the row for two degrees of freedom. The rightmost column is for significance of .001. We could thus conclude, "It is extremely unlikely that this sample of sites was selected from a population in which sites were evenly distributed across environmental settings ($\chi^2 = 13.8902$, $p < .001$)." Or, "The difference between our survey results and the expected results was very highly significant ($\chi^2 = 13.8902$, $p < .001$)."

PRACTICE

1. You have made surface collections at the Granger and Rawlins sites. Both collections include the same kinds of pottery, and you want to investigate whether the two sites differ in regard to the proportions of different pottery types. At the Granger site, you collected 162 sherds of the type Serengeti Plain, 49 sherds of the type Mandarin Orange, and 57 sherds of the type Zane Gray; from the Rawlins site you have 40 sherds of Serengeti Plain, 43 sherds of Mandarin Orange, and 49 sherds of Zane Gray. After considering possible sampling biases, you decide to use the collections as random samples from the populations consisting of all the sherds in each site. Estimate the proportions of the three pottery types at each site. Draw a bullet graph comparing the estimated proportions for the two sites with error bars for the 80%, 95%, and 99% confidence levels. (Think carefully about how to arrange the graph so that the error ranges you want to compare to each other are most easily compared.) How confident are you that the two sites differ in

Table 13.10. Temper and Surface Finish for Sherds from the Opelousas Site

Temper	Surface	Temper	Surface	Temper	Surface
sand	red	shell	plain	shell	red
sand	red	sand	plain	shell	red
sand	red	shell	red	sand	plain
shell	plain	shell	plain	sand	plain
sand	red	sand	red	sand	red
sand	plain	shell	plain	sand	red
sand	red	shell	red	sand	plain
shell	plain	sand	red.	sand	red
shell	red	shell	red	shell	plain
shell	red	shell	red	sand	red
sand	plain	sand	plain	sand	red
sand	red	sand	red	sand	red
sand	red	shell	plain	shell	red
sand	plain	shell	red	shell	plain
sand	plain	sand	red	shell	red
shell	plain	sand	plain	shell	plain
shell	plain	sand	plain		
shell	red	shell	plain		

regard to proportions of ceramic types? Summarize the conclusions of your graphical comparison in one or two sentences.

2. Approach the issues raised in Question 1 by evaluating the strength and significance of the association between the variables site and pottery type. Summarize your results in one sentence. How do these results compare with those obtained in Question 1? What are the advantages and disadvantages of approaching these issues with χ^2 rather than by estimating population proportions?

3. From the Opelousas site you have recovered a pitifully small collection of eroded sherds. You can't tell much about them except that some are tempered with shell and some with sand and that some were finished with a red slip while others have plain surfaces. The complete data are given in Table 13.10. Investigate the statistical significance and strength of any association between temper material and surface finish with the sample that you have. Summarize the meaning of your results in one clearly worded sentence.

Chapter *14*

Relating a Measurement Variable to Another Measurement Variable

In Chapters 11 and 12 we investigated the relationship between a measurement variable and a categorical variable. We took two approaches to this task. The first was to estimate population means for the measurement variable in each of the categories of the categorical variable and attach error ranges to those estimates. The second approach was to use either a two-sample *t* test (if only two categories were involved) or an analysis of variance (if more than two categories were involved). In Chapter 13 we investigated the relationship between two categorical variables. Once again we took two approaches. The first was to estimate population proportions for one of the variables in each of the categories of the other variable and attach error ranges to those estimates. The second approach was to use a chi-square test to evaluate significance and Cramer's *V* to evaluate strength of association. There remains only to investigate the relationship between two measurement variables to complete all the possible combinations, and that is the subject of

Table 14.1. Observations of Site Area and Number of Hoes in Collections of 100 Artifacts Made at Oasis Phase Sites in the Río Seco Valley

Site area (ha)	Number of hoes per 100 artifacts	Site area (ha)	Number of hoes per 100 artifacts	Site area (ha)	Number of hoes per 100 artifacts
19.0	15	14.0	19	10.9	31
16.4	14	13.0	16	9.6	39
15.8	18	12.7	22	16.2	23
15.2	15	12.0	12	7.2	36
14.2	20	11.3	22		

this chapter. We will see that one approach here is so powerful that we will not really consider alternative approaches.

Table 14.1 provides an example set of data consisting of observations on 14 known sites of the Oasis phase in the Río Seco valley. At each of the sites a systematic program of surface collection was undertaken to produce a sample of exactly 100 artifacts. After careful consideration of sources of bias we decide that we are willing to work with this sample of sites as if it were a random sample. Similarly considering sources of bias for the artifact collections, we decide we are willing to treat each as if it were a random sample of artifacts on the surface. Since each collection consists of 100 artifacts, the number of hoes in each is the percentage of hoes in the collection, and simultaneously our best approximation of the percentage of hoes in each population (that is, the population of artifacts on the surface at each of the sites). In effect, of course, what we are dealing with here is a percentage, and percentages like this make perfectly suitable measurement variables to study in this way. What we want to investigate here is whether there is any relationship between the area of the site, as indicated by the extent of artifacts visible on the surface and the number of hoes collected in the 100-artifact sample.

LOOKING AT THE BROAD PICTURE

As usual, drawing a plot that presents a picture of important aspects of the patterns to be observed is a good way to begin. The relationship between two measurement variables is best illustrated by a *scatter plot* (Figure 14.1). Each x in the scatter plot represents one of the sites, and its position is determined according to the area of the site (in the horizontal direction) and the number of hoes in the collection of 100 artifacts (in the vertical direction).

Simple observation of this scatter plot begins to reveal something of the relationship between these two variables. The points toward the left of the graph (that is, with low values for site area) tend to fall fairly high on the

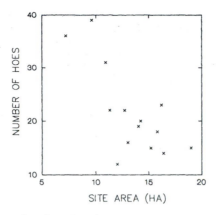

Figure 14.1. Scatter plot of number of hoes per 100 artifacts collected versus site area.

graph (that is, they have high values for number of hoes). The points farther toward the right of the graph (that is, with high values for site area) tend to fall fairly low on the graph (that is, they have low values for hoes). This suggests a pattern of larger sites having relatively fewer hoes per 100 artifacts and smaller sites having relatively more hoes per 100 artifacts.

In looking for patterns in scatter plots, especially if they are not very clear, it may sometimes help to look at groups of points separately and think of the levels in these subbatches. For example, look at the points representing small sites (between 5 and 10 ha) in Figure 14.1. There are only two such small sites, and both points fall very high on the graph, indicating that both sites have very high numbers of hoes. The center of this small batch of two sites is clearly quite high, perhaps around 37 hoes. In fact, these two smallest sites have the largest numbers of hoes of all the sites. Next look at the points in Figure 14.1 that represent the middle-sized sites (between 10 and 15 ha). All these points fall lower on the graph than the points representing the small sites. The center of this batch of middle-sized sites falls lower than the center for the small sites, probably somewhere near 20 hoes. Clearly the middle-sized sites have fewer hoes than the small sites. Finally, look at the points representing the large sites (between 15 and 20 ha). The center of this batch is lower still, perhaps around 15 hoes. The same pattern emerges from this more detailed examination of the scatter plot that we saw on simple inspection: in general, the bigger the site, the smaller the number of hoes per 100 artifacts.

This detailed way of looking at the scatter plot suggests one way we could approach this problem. We could treat site area as three categories (small, medium, and large) and estimate the mean number of hoes per 100 artifacts in each of these categories. Then we could attach error ranges to these estimates and draw a bullet graph to illustrate the overall patterns. Or we could perform an analysis of variance—the other technique applicable to investigating the relationship between a measurement and a set of categories. Measurement variables can always be converted into a set of categories in this way, and sometimes it is useful to do so. There is, however, a much more powerful way to approach the investigation of the relationship between two measurements.

LINEAR RELATIONSHIPS

The easiest kind of relationship to describe between two measurements is a *linear*, or straight-line, relationship. Such a relationship is called linear because it is represented by a straight line on a scatter plot. Perhaps the simplest possible relationship between two measurements is when the one equals the other. If we let X represent one of the measurements and Y represent the other, then the relationship of equivalence is simply expressed by the equation $Y = X$. For any given value of X, there is a corresponding value of Y, which is determined easily by the equation. For example, when $X = 5$, then $Y = 5$; when $X = -10$, then $Y = -10$. The values of X and Y are plotted on the graph in Figure 14.2. (By convention we always use the horizontal axis for X and the vertical axis for Y.) All the points representing pairs of X and Y values that satisfy the equation $Y = X$ lie along the line labeled $Y = X$ in Figure 14.2—a perfect straight-line relationship between X and Y.

The other lines in Figure 14.2 also represent perfect straight-line relationships between X and Y. They are labeled with the corresponding equations. The positions of these lines can be determined experimentally. For example, the line that represents $Y = -2X$ is defined by all the points that satisfy the equation $Y = -2X$. These include $X = 5$, $Y = -10$; $X = -7$, $Y = 14$ and so on. The equations in Figure 14.2 are algebraic expressions of relationships between two measurements, and the lines on the graph are geometric expressions of the same relationships. Each equation is a complete description of the corresponding line. If this is at all unclear, it is a good idea to experiment on your own with some equations and their corresponding graphs. Make up some values for X, calculate the corresponding values for Y, and plot the points.

Comparison of the equations in Figure 14.2 reveals one property of the relationship between equations and lines. If Y simply equals X multiplied by some number, then the relationship is represented geometrically by a straight line, and that straight line passes through the origin of the graph. (The origin is the point where $X = 0$ and $Y = 0$.) The number by which X is multiplied in the equation is called the *coefficient* of X, and it is this coefficient that gov-

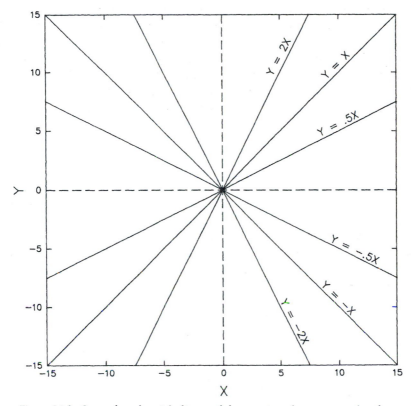

Figure 14.2. Some plotted straight lines and the equations that correspond to them.

erns the *slope* of the line. If the coefficient of X is positive, then the line rises as it moves from left to right. If the coefficient is negative, then the line falls as it moves from left to right. The larger the absolute value of the coefficient, the steeper the slope. That is, $Y = 2X$ has a steeper slope than $Y = .5X$, and $Y = -2X$ has a steeper slope than $Y = -.5X$. (In the equation $Y = X$, of course, the coefficient of X is 1.)

Figure 14.3 illustrates the other principal characteristic of straight lines on a graph—their positions relative to the origin. All the lines on the graph in Figure 14.3 have the same slope—the coefficient of X is .5 in every case. They differ, however, in the degree to which they are offset from the origin. These equations differ only in having an additional term added to the product of X and its coefficient. The line corresponding to $Y = .5X + 5$ crosses the Y axis at the point where $Y = 5$. (This, of course, has to be true because X is 0 at the Y axis, and when $X = 0$ then $Y = 5$.) This additional term is called the Y

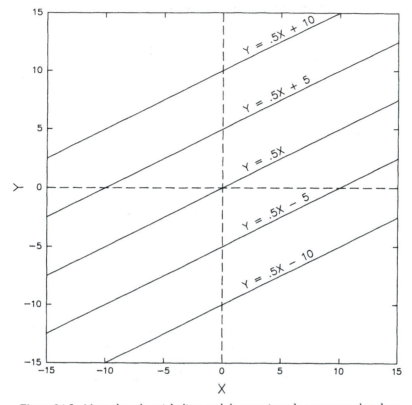

Figure 14.3. More plotted straight lines and the equations that correspond to them.

intercept since it is the value of Y when X is 0, which is to say, the value of Y where the straight line crosses the Y axis.

Thus for any straight-line relationship between X and Y we can write an equation in the form

$$Y = bX + a$$

where b is the slope of the line, and a is the Y intercept, or value of Y where the line crosses the Y axis.

This specifies exactly what the relationship between X and Y is. It enables us to say, for a given value of X, what Y will be. By convention, we always take X as a given and let Y's value depend on X. Thus X is the *independent variable* and Y is the *dependent variable*.

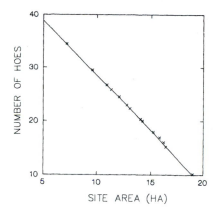

Figure 14.4. If the scatter plot in Figure 14.1 had looked like this, it would have been easy to fit a straight line to the points.

THE BEST-FIT STRAIGHT LINE

We have strayed rather far from the example where we wanted to investigate the relationship between site area and number of hoes collected per 100 artifacts. The point of the discussion of straight-line relationships, however, was to make clear exactly what kind of mathematical relationship we might expect to find between these two measurements. If the relationship between site area and number of hoes collected can be described reasonably accurately as a straight-line relationship, then we can characterize it in these terms. If, for example, the scatter plot in Figure 14.1 had looked like Figure 14.4 instead, we would find it quite easy to apply the principles of straight-line equations just discussed. The points in Figure 14.4 do fall almost perfectly along a straight line, and an approximation of that line has been drawn on the graph. We could measure the slope of the line and determine the Y value of the point at which it crosses the Y axis and write an equation that specified the relationship between the two measurements algebraically.

The problem, of course, is that the points in the real scatter plot from our example data did not fall almost perfectly along a straight line. While the general pattern of declining numbers of hoes per 100 artifacts with increasing site area was clear, no straight line could be drawn through all the points. There are so many advantages to working with straight-line relationships, though, that it is worth trying to draw a straight line on Figure 14.1 that represents the general trend of the points as accurately as possible—a *best-fit straight line*. The statistical technique for accomplishing this is *linear regression*.

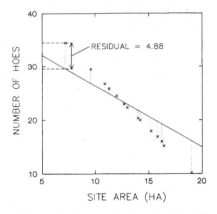

Figure 14.5. A straight line that does *not* fit the points from Figure 14.4 very well.

The conceptual starting point for linear regression is to think exactly what criterion would determine which line, of all the possible straight lines we could draw on the scatter plot, would fit the points best. Clearly, we would like as many of the points as possible to lie as close to the line as possible. Since we take the values of X as given, we think of closeness to the line in terms of Y values only. That is, for a given X value, we think of how badly the point "misses" the line in the Y direction on the graph. These distances are called *residuals*, for reasons that will become clear later on.

We can explore the issue of residuals with the completely fictitious scatter plot of Figure 14.4. Since the points in this scatter plot do fall very closely along a straight line, it is a bit easier to see good and bad fits. Figure 14.5 illustrates a straight line that does not fit the pattern of points nearly so well as the one in Figure 14.4. We can see that simply by inspection. We could put a finer point on just how bad the fit is by measuring the residuals, which are indicated with dotted lines in Figure 14.5. The measurements, of course, would be taken vertically on the graph (that is, in the Y direction) and would be in terms of Y units (that is, in this example, in terms of numbers of hoes). The same operation could be performed algebraically as well. Since the line corresponds to a specific linear equation relating X and Y, we could use that equation to calculate, for each X value, the value of Y that would be "correct" according to the relationship the line represents. The difference between that "correct" value of Y and the actual value of Y would correspond to the graphical measurement of the residual. The residual and its measurement are shown for the leftmost point on the scatter plot in Figure 14.5. This point falls 4.88 Y units above the straight line. The residual corresponding to this point, then, is 4.88. This means that this site actually had

4.88 more hoes per 100 artifacts collected than the value we would calculate *based on the straight line drawn in Figure 14.5.*

We can easily see that the straight line in Figure 14.5 could be adjusted so that it followed the trend of the points better by twisting it around a bit in a clockwise direction. If we did this, the dotted lines representing the residuals could all be shortened substantially. Indeed, we would put the line back the way it was in Figure 14.4 and the residuals would all be zero or nearly zero. Thus we can see that minimizing residuals provides a mathematical criterion that corresponds well to what makes good sense to us from looking at the scatter plot. The better the fit between the straight line and the points in the scatter plot, the smaller the residuals are collectively.

The residuals amount to deviations between two alternate values of *Y* for a given value of *X*. There is the *Y* value represented by the straight line and there is the *Y* value represented by the data point. As usual in statistics, it turns out to be most useful to work not directly with these deviations but with the squares of the deviations. Thus, the most useful mathematical criterion is that *the best-fit straight line is the one for which the sum of the squares of all the residuals is least.* From this definition comes a longer name for the kind of analysis we are in the midst of: *least-squares regression.*

The core of the mathematical complexity of regression analysis, as might be expected, concerns how we determine exactly which of all the possible straight lines we might draw provides the best fit. Fortunately, it is not necessary to approach this question through trial and error. Let's return to the general form of the equation for a straight line relating *X* and *Y*:

$$Y = bX + a$$

It can be shown mathematically that the following two equations produce values of *a* and *b* that, when inserted in the general equation, describe the best-fit straight line:

$$b = \frac{n \, \Sigma \, X_i \, Y_i - (\Sigma \, Y_i) \, (\Sigma \, X_i)}{n \, \Sigma \, X_i^2 - (\Sigma \, X_i)^2}$$

and

$$a = \bar{Y} - b\bar{X}$$

where

$n =$ the number of elements in the sample;

$X_i =$ the X value for the ith element;
$Y_i =$ the Y value for the ith element.

Since the summations involved in the equation for b are complex, it is perhaps worth explaining the operation in detail. For the first term in the numerator of the fraction, $n \Sigma X_i Y_i$, we multiply the X value for each element in the sample by the Y value for the same element, then sum up these n products, and multiply the total by n. For the second term in the numerator, $(\Sigma Y_i)(\Sigma X_i)$, we sum up all n X values, sum up all n Y values, and multiply the two totals together. For the first term in the denominator, $n \Sigma X_i^2$, we square each X value, sum up all these squares, and multiply the total by n. And for the second term in the denominator, $(\Sigma X_i)^2$, we sum up all the X values and then square the total. Having arrived at a value for b, deriving the value of a is quite easy by comparison. We simply subtract the product of b times the mean of X from the mean of Y.

There are computational shortcuts for performing these cumbersome calculations, but, in fact, there is little likelihood that any reader of this book will perform a regression analysis without a computer, so we will not take up space with these shortcut calculations. Neither will we laboriously work these equations through by hand to arrive at the actual numbers for our example. This example has been performed the way everyone now can fully expect to perform a regression analysis—by computer. The point of including the equations here, then, is not to provide a means of calculation but instead to provide insight into what is being calculated and thus into what the results may mean.

PREDICTION

Once we have the values of a and b, of course, we can specify the equation relating X and Y and, by plugging in any two numbers as given X values, determine the corresponding Y values and use these two points to draw the best-fit straight line on the graph. If we do this for the data from Table 14.1, we get the result shown in Figure 14.6. The values obtained in this regression analysis are

$$a = 47.802$$

and

$$b = -1.959$$

Thus the equation relating X to Y is

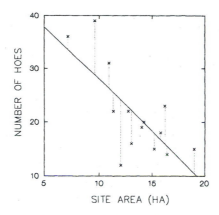

Figure 14.6. The best-fit straight line for the points from Figure 14.1.

$$Y = -1.959X + 47.802$$

or

$$Number\ of\ hoes = (-1.959 \times site\ area) + 47.802$$

This equation literally enables us to "predict" how many hoes there will be per 100 artifacts collected if we know the site area. For example, if the site area is 15.2 ha, we predict

$$Y = (-1.959)\ (15.2) + 47.802 = 18.03$$

Thus, if the relationship between X and Y described by the regression equation holds true, a site with an area of 15.2 ha should yield 18.03 hoes in a collection of 100 artifacts. There actually was a site in the original data set with an area of 15.2 ha, and the collection of 100 artifacts from that site had 15 hoes. For this site, then, reality fell short of the predicted number of hoes by 3.03. Thus the residual for that site is –3.03, representing a bit of variation unpredicted or "unexplained" by the regression equation. (The name "residual" is used because residuals represent unexplained or leftover variation.) The prediction based on the regression equation is, however, a better prediction than we would otherwise be able to make. Without the regression analysis our best way to predict how many hoes would be collected at each site would be to use the mean number of hoes for all sites, or 21.57 hoes. This would have meant an error of 6.57 hoes in the case of the 15.2 ha site. In this instance, then, the regression equation has enabled us to predict how many hoes would be found on the basis of site area more accurately than we could if we were unaware of this relationship. This will not necessarily be true for every single case in a regression analysis, but it will be true on average.

Regression analysis, then, has helped us to predict or "explain" some of the variation in number of hoes collected per 100 artifacts. It has, however, still left some of the variation "unexplained." We do not know why the 15.2 ha site had 3.03 fewer hoes than we expected. The residuals represent this unexplained variation, a subject to which we shall return below.

HOW GOOD IS THE BEST FIT?

We know that the equation

$$Y = (-1.959) X + 47.802$$

represents the best-fit straight line for our example data, so that the sum of the squares of the residuals is the lowest possible (for straight-line equations). These residuals are shown with dotted lines in Figure 14.6. We notice immediately that some of them are quite large. Although the best-fit straight line does help us predict or explain some of the variability in number of hoes collected, it clearly does not fit the data as well as we might have hoped. It would be useful for us to be able to say just how good a fit it is, and the very process of determining the best-fit straight line provides us with a way to do so. Since the best-fit straight line is the one for which the sum of the squares of the residuals is least, then the lower the sum of the squares of the residuals, the better the fit. The sum of the squares of the residuals becomes a measure of how well the best-fit straight line fits the points in the scatter plot.

The sum of the squares of the residuals, of course, can never be less than 0, because there will never be a negative number among the squared residuals that are summed. (Even negative residuals have positive squares.) The sum of the squares of the residuals will only be 0 when all the residuals are 0. This only happens when all the points lie exactly on the straight line and the fit is thus perfect. There is no fixed upper limit on the sum of the squares of the residuals, however, because it depends on the actual values taken by Y. It would be useful if we could determine this upper limit because then we would know just where, between the minimum and maximum possible values, a particular sum of squared residuals lay. We could then determine whether the best-fit straight line really was closer to the best of all possible fits (a value of zero for the sum of the squares of the residuals) or the worst of all possible fits (whatever that maximum value for the sum of the squares of the residuals might be). It turns out that the maximum value the sum of the squares of the residuals can have is the sum of the squares of the deviations of Y from its mean. (The sum of the squares of the deviations of Y from its mean is, of course, the numerator in the calculation of the variance of Y—that is, $\Sigma (y_i - \overline{Y})^2$.) Thus the ratio

$$\frac{(sum\ of\ the\ squares\ of\ residuals)}{\Sigma(y_i - \overline{Y})^2}$$

ranges from zero to one. Its minimum value of zero indicates a perfect fit for the best-fit straight line because it only occurs when all the residuals are zero. Its maximum value of one indicates the worst possible fit because it occurs when the sum of the squares of the residuals is as large as it can be for a given set of values of Y (that is, equal to $\Sigma (y_i - \overline{Y})^2$).

This ratio, then, enables us to say, on a scale of zero to one, how good a fit the best-fit straight line is. Zero means a perfect fit, and one means the worst possible fit. It is easier intuitively to use a scale on which one is best and zero is worst, so we customarily reverse the scale provided by this ratio by subtracting the ratio from one. (If this does not make intuitive good sense to you, try it with some numbers. For example, .2 on a scale from zero to one becomes .8 on a scale from one to zero.) This ratio, when subtracted from one, is called r^2, and

$$r^2 = 1 - \frac{(sum\ of\ the\ squares\ of\ residuals)}{\Sigma (y_i - \overline{Y})^2}$$

The ratio, r^2, amounts to a ratio of variances. The denominator is the original variance in Y (omitting only the step of dividing by $n - 1$) and the numerator is the variance that Y has from the best-fit straight line (again omitting only the step of dividing by $n - 1$). Including the step of dividing by $n - 1$ would have no effect on the result since it would occur symmetrically in both numerator and denominator.

If the variation from the best-fit straight line is much less than the original variation of Y from its mean, then the value of r^2 is large (approaching one) and the best-fit straight line is a good fit indeed. If the variation from the best-fit straight line is almost as large as the original variation of Y from its mean, then the value of r^2 is small (approaching zero) and the best-fit straight line is not a very good fit at all. Following from this logic, it is common to regard r^2 as a measure of the *proportion of the total variation in Y explained by the regression*. This also follows from our consideration of the residuals as variation unexplained or unpredicted by the regression equation. All this, of course, amounts to a rather narrow mathematical definition of "explaining variation," but it is useful nonetheless within the constraints of linear regression. For our example, r^2 turns out to be .535, meaning that 53.5% of the variation in number of hoes per collection of 100 artifacts is explained or accounted for by site area. This is quite a respectable amount of variation to account for in this way.

More commonly used than r^2 is its square root, r, which is also known as *Pearson's r*, or the *product-moment correlation coefficient*, or just the *correlation coefficient*. We speak, then, of the *correlation* between two measurement variables as a measure of how good a fit the best-fit straight line is. Since r^2 ranges from zero to one, its square root must also range from zero to one. While r^2 must always be positive (squares of anything always are), r can be either positive or negative. We give r the same sign as b, the slope of the best-fit straight line. As a consequence, a positive value of r corresponds to a best-fit straight line with a positive slope and thus to a positive relationship between X and Y, that is, a relationship in which as X increases Y also increases. A negative value of r corresponds to a best-fit straight line with a negative slope and thus to a negative relationship between X and Y, that is, a relationship in which as X increases Y decreases. The correlation coefficient r, then, indicates the direction of the relationship between X and Y by its sign, and it indicates the strength of the relationship between X and Y by its absolute value on a scale from zero for no relationship to one for a perfect relationship (the strongest possible). In our example, $r = -.731$, which represents a relatively strong (although negative) correlation.

SIGNIFICANCE AND CONFIDENCE

Curiously enough, the question of significance has not arisen up to now in Chapter 14. The logic of our approach to relating two measurement variables has been very different from our approach to relating two categorical variables or one measurement variable and one categorical variable. Through linear regression, however, we have arrived at a measure of strength of the relationship, r, the correlation coefficient. This measure of strength is analogous to V, the measure of strength of association between two categorical variables. It is analogous to the actual differences between means of subgroups in analysis of variance as an indication of the strength of relation between the dependent and independent variables. We still lack, however, a measure of the significance of the relationship between two measurements. What we seek is a statistic analogous to χ^2 for two categorical variables, or t or F for a categorical variable and a measurement—a statistic whose value can be translated into a statement of how likely it is that the relation we observe is no more than the effect of the vagaries of sampling.

Much of our discussion about arriving at the best-fit straight line and providing an index of how good a fit it is centered on variances and ratios of variances. This sounds a great deal like analysis of variance, and indeed it is by calculating F as a ratio of variances that we arrive at the significance level in a regression analysis. In analysis of variance we had

$$F = \frac{s_B^2}{s_W^2} = \frac{SS_B / d.f.}{SS_W / d.f.}$$

which is to say

$$F = \frac{(sum\ of\ squares\ between\ groups\ /\ d.f.)}{(sum\ of\ squares\ within\ groups\ /\ d.f.)}$$

In regression analysis we have

$$F = \frac{(sum\ of\ squares\ explained\ by\ regression\ /\ d.f.)}{(sum\ of\ squares\ unexplained\ by\ regression\ /\ d.f.)}$$

This is equivalent to

$$F = \frac{r^2 / 1}{(1 - r^2) / (n - 2)}$$

In our example, $F = 13.811$, with an associated probability of .003. As usual, very low values of p in significance tests indicate very significant results. There are several ways to think about the probability values in this significance test. Perhaps the clearest is that this result indicates a probability of .003 of selecting a random sample with a correlation this strong from a population in which these two variables were unrelated. That is, there are only 3 chances in 1000 that we could select a sample of 14 sites showing this strong a relation between area and number of hoes from a population of sites in which there was no relation between area and number of hoes. Put yet another way, there is only a 0.3% chance that the relationship we observe in our sample between site area and number of hoes reflects nothing more than the vagaries of sampling. If we are willing to treat these 14 sites as a random sample of Oasis phase sites from the Río Seco valley, then, we are 99.7% confident in asserting that larger Oasis phase sites in the Río Seco valley tend to have fewer hoes per 100 artifacts on their surfaces.

As usual, significance probabilities can be used to tell us how likely it is that the observation of interest in our sample (in this case the relationship between site area and number of hoes) does not actually exist in the popula-

---

Be Careful How You Say It

We might report the results of the example regression analysis in the text by saying, "For Oasis phase sites in the Río Seco valley there is a moderately strong correlation between site area (X) and number of hoes per collection of 100 artifacts (Y) ($r = -.731$, $p = .003$, $Y = -1.959X + 47.802$)." This makes clear what relationship was investigated; it lets the reader know what significance test was used; it provides the results in terms of both strength and significance; and it states exactly what the best-fit linear relationship is. Like "significance," the word "correlation" has a special meaning in statistics that differs from its colloquial use. It refers specifically to Pearson's r and other analogous indexes of the relationship between two measurements. Just as "significant" should not be used in statistical context to mean "important" or "meaningful," "correlated" should not be used in statistical context to refer simply to a general correspondence between two things.

---

tion from which the sample was selected. We can also discuss regression relationships in terms of confidence, in a manner parallel to our earlier use of error ranges for different confidence levels. In this case, instead of an individual estimate ± an error range, it is useful to think of just what the relationship between the two variables is likely to be in the population from which our sample came. We know from the significance probability obtained in our example that it is extremely unlikely that there is no relationship at all between site area and number of hoes in the population of sites from which our 14 sites are a sample. The specific relationship expressed by the regression equation derived from analysis of our sample is our best approximation of what the relationship between site area and number of hoes is in the population. But, as in all our previous experience with samples, the specific relationship observed in the sample may well not be exactly the same as the specific relationship that exists in the population as a whole. Most likely the regression equation we would obtain from observing the entire population (if we could) would be similar to the one we have derived from analysis of the sample. It is less likely (but still possible) that the relationship in the population as a whole is rather different from the relationship observed in the sample. And, as the significance probability has already told us, it is extremely unlikely (only 3 chances in 1000) that there is no relationship at all between site area and hoes in the population.

This range of possible relationships that might exist in the population our sample came from, and their varying probabilities can be depicted graphically as in Figure 14.7. It is neither very practical nor very enlightening to discuss the calculation of the curves that delimit this 95% confidence

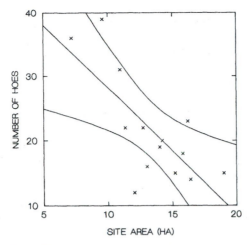

Figure 14.7. The best-fit straight line with its 95% confidence region.

region. In practice, it is almost unimaginable now to produce such a graph except by computer, so we will concentrate on what the graph tells us. The 95% confidence region, which includes the best-fit straight line for our sample in its center, depicts the zone within which we have 95% confidence that the best-fit straight line for the population lies. There is only a 5% chance that the best-fit straight line for the population of sites from which our sample of 14 was selected (if we could observe the entire population) would not lie entirely between the two curves. Determination of this confidence region, then, enables us to think usefully about the range of possible relationships between site area and number of hoes likely to exist in the population from which our sample came. Depiction of this confidence region relates to the significance probability in the same way that our previous use of error ranges for different confidence levels related to parallel significance tests.

ANALYSIS OF RESIDUALS

The regression analysis described in the example used above has enabled us to explain a portion of the variation in number of hoes per collection of 100 artifacts. One possible interpretation of these results is that larger settlements contained larger numbers of craft workers and elite residents and fewer farmers. Thus hoes were scarcer in the artifact assemblages at the larger sites. (We would presumably have had something like this in mind in the first place or we would not likely have been interested in investigating the

Table 14.2. Hoes at Oasis Phase Sites in the Río Seco Valley: Predictions and Residuals

Site area (ha)	Number of hoes per 100 artifacts	Number of hoes predicted by regression on site area	Residual number of hoes
19.0	15	10.59	4.41
16.4	14	15.68	−1.68
15.8	18	16.86	1.14
15.2	15	18.03	−3.03
14.2	20	19.99	0.01
14.0	19	20.38	−1.38
13.0	16	22.34	−6.34
12.7	22	22.93	−0.93
12.0	12	24.30	−12.30
11.3	22	25.67	−3.67
10.9	31	26.45	4.55
9.6	39	29.00	10.00
16.2	23	16.07	6.93
7.2	36	33.70	2.30

relationship between site area and number of hoes at all. We would also presumably have provided the additional evidence and argumentation necessary to make this a truly convincing interpretation.)

Since the regression analysis has explained part of the variation in number of hoes, it has also left another part of this variation unexplained. This unexplained variability is made specific in the form of the residuals. The 15.2-ha site that we discussed, for instance, actually had 3.03 fewer hoes than the regression analysis led us to expect, based on the size of the site. This 3.03 is its residual, or leftover variation. For each site there is, likewise, a residual representing how much the observed number of hoes differed from the predicted number of hoes. Table 14.2 provides the original data together with two new items. For each site, the number of hoes per collection of 100 artifacts predicted on the basis of the regression equation relating number of hoes to site area is listed. Then comes the residual for each site (that is, the number of hoes actually collected minus the number predicted by the regression equation).

In examining the residuals, we note as expected that some sites had considerably fewer hoes than we predicted and some had substantially more than we predicted. We can treat these residuals as another variable whose relationships can be explored. In effect, the regression analysis has created a new measurement—the variation in number of hoes unexplained by site size. We can deal with this new measurement just as we would deal with any measurement we might make. We would begin to explore it by looking at a stem-and-leaf plot and perhaps a box-and-dot plot. We would be interested,

---

Statpacks

Regression analysis is hardly ever performed any more except by computer. Different statpacks use a variety of vocabularies to talk about it, in part because linear regression is only the tip of the iceberg. Regression analysis is really a whole family of analytical approaches involving curved line fitting in addition to straight line fitting and incorporating a number of variables simultaneously instead of just two. Any very large and powerful statpack will perform many of these other kinds of analysis as well, and the simple, but powerful, linear regression techniques discussed here may be embedded in this broader family of analyses. Consequently, the commands or menu selections that produce a simple linear regression vary substantially from one statpack to another and are often much more complicated than it seems like they need to be. Recourse to the manual for your particular program is likely to be necessary. Some statpacks integrate scatter plots into the procedures that perform regression analysis as an option, while others perform the numerical analysis as one operation and produce scatter plots as a different operation. Usually the inclusion of the curves delimiting a confidence region for the best-fit straight line is an option to be specified as part of the production of a scatterplot. Residuals, of course, are calculated as part of the regression analysis, but to be able to use them as a new measurement and pursue further analysis with them it is usually necessary to save them by specifying this as an option to the regression analysis. Typically this results in the creation of a new data file in the normal format your statpack uses for data files. The new file will have the same cases as the original data file and a variable whose values are the residuals from the regression analysis.

---

for example, in the possibility of multiple peaks in this new batch of numbers. A two-peaked shape would suggest two distinct sets of sites, probably one with substantially more hoes than we would expect (given site size), and one with substantially fewer hoes than we would expect. We might be able to determine some other characteristics of these two groups of sites that helped us to understand why they deviated in such different ways from the number of hoes we would expect, given their size. If the shape is single-peaked we might go on to explore the relationship between this new batch of measurements and other variables. For example, we might imagine that, in addition to site area reflecting the presence of nonfarming specialists, residents of sites in very fertile soils might dedicate themselves more intensively to farming than residents of sites in very poor soils. We might, then, investigate the relationship between our new measurement (the residuals from the regression analysis) and fertility of soils for each site. In short, we could do anything with this new measurement that we could do with any measurement and pursue our understanding beyond what we learned about the relationship between site area and number of hoes. There are also ways of incorporating a number of variables simultaneously into a single regression analysis, which accomplishes a similar result, but such *multivariate analysis* is beyond the scope of this book.

ASSUMPTIONS AND ROBUST METHODS

It may come as a surprise that linear regression is not based on the assumption that both measurements involved have normal shapes. The shape assumptions that we must be alert to in linear regression have to do with the shapes of point distributions in scatter plots. Just as we examine stem-and-leaf plots to check for the single peak and symmetry that characterize a normal shape, we examine a scatter plot prior to linear regression for the shape of point distributions. What we need to see is a cloud of points of roughly oval shape. There should be no extreme outliers from the cloud, the oval should be of similar thickness throughout, and there should be no tendencies toward curvature of the whole oval. These three potential problems can be discussed separately.

First, outliers present severe risks to linear regression. Figure 14.8 provides an extreme example that should make the principle intuitively clear. The points in the lower left corner of the scatter plot clearly show an extremely strong negative correlation. The single outlier to the upper right, however, will cause the best-fit straight line to be as shown—a positive

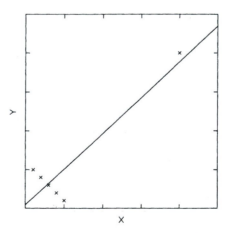

Figure 14.8. The devastating effect of a single outlier on the best-fit straight line.

correlation of some strength. Outliers have such a strong effect on the best-fit straight line that they simply cannot be overlooked. When outliers are identified, those cases should be examined with great care to see whether there is a measurement or data-recording error that can be corrected or whether there is some other reason to justify excluding them from the sample.

Second, oval shapes of points with very thin sections (or even worse, two or more separate oval clouds) are the equivalent of multipeaked shapes for single batches of numbers. They can create the same kinds of problems in linear regression that outliers do. Figure 14.9 shows another extreme example, where two ovals of points showing negative correlations of some strength turn into a single best-fit straight line with a positive slope when improperly analyzed together. Such a shape may occur in a scatter plot of two variables that, when looked at individually, have clearly single-peaked and symmetrical shapes. Shapes like this should be broken apart for separate analysis.

Third, tendencies toward curved patterns in the oval of points can prevent a very good fit of a straight line to a fundamentally linear pattern that just happens to be curved. There are ways to extend the logic of linear regression to more complex curvilinear relationships between variables, but it is usually much easier to straighten out the curve by transforming one or both variables. The kinds of transformations required are very like the transformations discussed in Chapter 5 and may be applied to either or both of the variables to remove tendencies toward curvature. As Figure 14.10 illustrates, if the scatter plot shows a tendency toward linear patterning but with the ends curving downward, a square root transformation of X will produce a straighter line. If stronger corrective action is called for, the logarithm of X

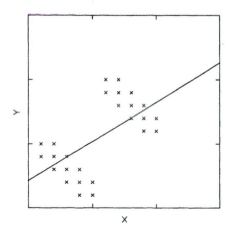

Figure 14.9. The effect of two oval clouds of points on the best-fit straight line.

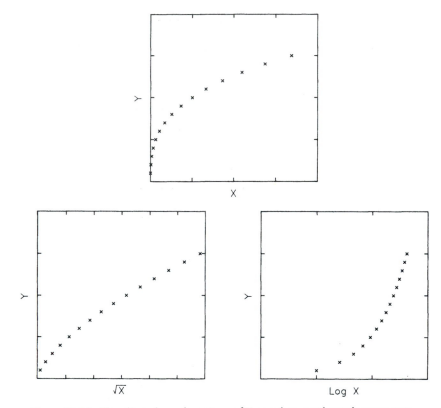

Figure 14.10. The effect of transformations of X on a downward curvilinear pattern.

can be used instead of the square root. Clearly, for the data in Figure 14.10, the logarithm of X is too strong a transformation, having produced just as curved a pattern in the opposite direction. Figure 14.11 illustrates transformations to correct linear patterns where the ends curve upward. For these data the square of X produces good results. Using the cube of X produces a

Table 14.3. Data from Storage Pits at Yenangyaung

Volume (m^3)	No. of artifacts	Volume (m^3)	No. of artifacts	Volume (m^3)	No. of artifacts
1.350	78	0.760	34	0.920	38
0.960	30	0.680	33	0.640	13
0.840	35	1.560	60	0.780	18
0.620	60	1.110	47	0.960	25
1.261	23	1.230	47	0.490	56
1.570	66	0.710	20	0.880	22
0.320	22	0.590	28		

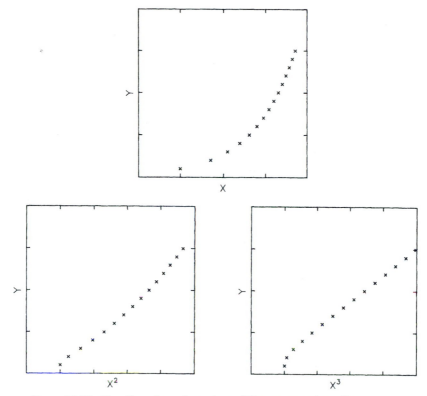

Figure 14.11. The effect of transformations of X on an upward curvilinear pattern.

stronger effect than is needed in this instance. Applying, for example, a square root transformation to X prior to analysis means, of course, that it is not X but rather $\sqrt{X}$ whose relationship to Y is being investigated. Thus it becomes $\sqrt{X}$ rather than X that is used in the regression equation to predict the values of Y.

PRACTICE

You have excavated a site near Yenangyaung that has a number of apparent storage pits containing artifacts and other debris. You wish to investigate whether the density of artifacts (the number per unit volume) is constant for all the pits. (Another way to phrase this is to ask yourself whether, knowing the volume of a pit, you could accurately predict the number of artifacts it contains.) The volume measurements and the number of artifacts recovered from complete excavation of each pit are given in Table 14.3.

1. Make a scatter plot of pit volume and number of artifacts. What does inspection of the scatter plot suggest about a relationship between them?

2. Perform a regression analysis for pit volume and number of artifacts. How can the relationship between number of artifacts and pit volume be expressed mathematically? How many artifacts would you expect to find in a pit whose volume was 1.000 m^3?

3. How much of the variation in number of artifacts is "explained" by pit volume? What is the statistical significance of the relationship between pit volume and number of artifacts? Produce a scatter plot showing the 90% confidence region for the best-fit straight line.

4. Sum up clearly and concisely what this regression analysis of the relationship between pit volume and number of artifacts has shown.

Chapter 15

Relating Ranks

Sometimes we have variables that at first glance appear to be measurements, but on further examination reveal themselves to be something less than actual measurements along a scale. Often they really amount to relative rankings rather than true measurements. For example, soil productivity is sometimes rated by producing an index with an arbitrary formula using such values as content of various nutrients, soil depth, capacity for water retention, and other variables that affect soil productivity. The formulas used in these ratings are carefully considered to produce a set of numbers such that we are sure that higher numbers represent more productive soils and lower numbers represent less productive soils. Such scales, for example, would allow us to say that a rating of 8 means more productive soils than a rating of 4. They seldom, however, leave us in position to say that a rating of 8 means soils twice as productive as a rating of 4. It is our inability to make this last statement that keeps such ratings from being true measurements. Instead, they are *rankings*. Rankings allow us to put things in rank order (most productive soil, second most productive soil, third most productive soil, etc.) but not to say *how much more* a high-ranking thing is than a low-ranking thing.

The logic of linear regression relies on the measurement principle. (Think of the scatter plots and the regression equations. If X is twice as large it places the corresponding point twice as far over on the scatter plot. If X is twice as large it has twice the effect on the prediction of Y by way of the regression equation.) If X is actually only a ranking rather than a true measurement, then we should feel uncomfortable about using regression. Instead

227

of performing a linear regression and attempting to predict the actual value of Y from X, we might use a *rank order correlation coefficient* to assess the strength and significance of a rank order relationship.

A rank order relationship has nothing to do with the actual magnitude of the rankings for either variable studied, but rather only with the order of the rankings. If we rank order a batch of numbers according to the values for X and this rank order is exactly the same as the rank order of values for Y, then X and Y show a perfect positive rank order relationship. That is, the highest value for X is for the case that also has the highest value for Y; the second highest value for X is for the case that also has the second highest value for Y; and so on. A perfect negative rank order relationship means that the case with the highest value for X has the lowest value for Y; the case with the second highest value for X has the second lowest value for Y; and so on until the case with the lowest value for X has the highest value for Y.

We can imagine a rank order correlation coefficient that works like Pearson's r, so that a perfect positive rank order relationship is assigned a value of 1; a perfect negative rank order relationship is assigned a value of -1; and intermediate relationships are assigned values between 1 and -1, depending on the extent to which the relationships approach one or the other of these ideal situations. Several such coefficients exist. One of the most frequently used is *Spearman's rank correlation coefficient* (r_s).

CALCULATING SPEARMAN'S RANK CORRELATION

Table 15.1 contains data for soil productivity ratings for 17 different soil zones in the Konsankoro Plain. The Neolithic occupation consisted of a series of sedentary village sites of remarkably consistent size. We take the number of village sites in each soil zone divided by the total number of km^2 covered by that zone to indicate how densely the zone was occupied, and we wish to investigate whether more productive soil zones were more densely inhabited.

The first step in calculating Spearman's rank correlation is to determine the rank orderings of all the cases for each of the two variables (taken separately). These rank orderings are also given in Table 15.1. Ties frequently occur in the soil productivity ratings. That is, for example, soil zones H and K are ranked in the lowest productivity category (1). These two least productive soil zones should be rank ordered 1 and 2, but we have no basis for putting one above the other since they are tied in the productivity ratings. As a consequence, we assign each a rank order of 1.5 (the mean of 1 and 2). Soil zones C, I, and P are tied with productivity ratings of 3. These would be soil zones 5, 6, and 7 in rank order if we could determine which to put above the other. Since we cannot make this determination, each is assigned a rank

Table 15.1. Soil Productivity and Villages in the Konsankoro Plain

Soil zone	Productivity rating X	No. of villages per km^2 Y	Rankings X	Y	d	d^2	t$_x$	T$_x$	t$_y$	T$_y$
A	2	.26	3.5	2	1.5	2.25	2	0.5	1	0.0
B	6	1.35	11.5	14	−2.5	6.25	2	0.5	1	0.0
C	3	.44	6	6	0.0	0.00	3	2.0	1	0.0
D	7	1.26	13.5	12	1.5	2.25	2	0.5	1	0.0
E	4	.35	8.5	4	4.5	20.25	2	0.5	1	0.0
F	8	2.30	16	17	−1.0	1.00	3	2.0	1	0.0
G	8	1.76	16	16	0.0	0.00	3	2.0	1	0.0
H	1	.31	1.5	3	−1.5	2.25	2	0.5	1	0.0
I	3	.37	6	5	1.0	1.00	3	2.0	1	0.0
J	5	.78	10	11	−1.0	1.00	1	0.0	1	0.0
K	1	.04	1.5	1	0.5	.25	2	0.5	1	0.0
L	8	1.62	16	15	1.0	1.00	3	2.0	1	0.0
M	7	1.34	13.5	13	0.5	.25	2	0.5	1	0.0
N	2	.47	3.5	7	−3.5	12.25	2	0.5	1	0.0
O	4	.56	8.5	9	−0.5	.25	2	0.5	1	0.0
P	3	.48	6	8	−2.0	4.00	3	2.0	1	0.0
Q	6	.76	11.5	10	1.5	2.25	2	0.5	1	0.0

$$\Sigma d^2 = 56.5 \qquad \Sigma T_x = 17.0 \qquad \Sigma T_y = 0.0$$

order of 6 (the mean of 5, 6, and 7). Such a treatment is accorded whenever there are ties. No ties occur in the number of villages per km^2 (which actually is a true measurement), so the rank ordering is simpler. It begins at 1 for soil zone K and continues through zones A, H, and so on to zone F, which ranks 17th because it has the highest number of village sites per km^2.

Subtracting the rank orderings for villages per km^2 (Y) from the rank orderings for soil productivity (X) gives us the difference between rankings d, which we then square and sum up to get Σd^2.

The last four columns in Table 15.1 concern a correction that must be made for ties. The value t for each soil zone is the total number of soil zones that are tied at that ranking. For example, soil zone A has a value of $t_x = 2$ because a total of two zones (A and N) are tied at its productivity rating of 2. Since there are no ties for number of villages per km^2, all the values of t_y are 1. For each t value for each of the two variables, a value of T is obtained as follows:

$$T = \frac{t^3 - t}{12}$$

The calculation of Spearman's rank correlation requires three sums from Table 15.1: Σd^2, ΣT_x, and ΣT_y. A sum of squares is calculated for each of the two variables:

$$\Sigma x^2 = \frac{n^3 - n}{12} - \Sigma T_x$$

where ΣT_x is from Table 15.1, and n is the number in the sample (17 in this example). Thus, for the example in Table 15.1,

$$\Sigma x^2 = \frac{17^3 - 17}{12} - 17.0 = 408 - 17 = 391$$

and

$$\Sigma y^2 = \frac{17^3 - 17}{12} - 0.0 = 408 - 0.0 = 408$$

Spearman's rank correlation, then, is given by the equation

$$r_S = \frac{\Sigma x^2 + \Sigma y^2 - \Sigma d^2}{2 \sqrt{\Sigma x^2 \, \Sigma y^2}}$$

For the example in Table 15.1, then,

$$r_S = \frac{391 + 408 - 56.5}{2 \sqrt{(391)(408)}} = \frac{742.5}{798.8} = .93$$

Spearman's rank order correlation coefficient, then, between soil productivity and number of villages per km^2 in the Konsankoro Plain is .93, indicating a strong positive correlation. (Values for r_S can be interpreted in much the same manner as those for Pearson's r, although the two cannot be compared directly. That is, a Spearman's r_S of .85 between two variables cannot be said to indicate a stronger correlation than a Pearson's r of .80 between two other variables.)

If there are no ties, then we can easily see that $\Sigma T = 0$ (as in the case of number of villages per km^2 in Table 15.1). If there are no ties for either variable, then, there is no need to go to the trouble of figuring t and T, and the entire equation for Spearman's rank correlation is considerably simplified:

$$r_S = 1 - \frac{6\,\Sigma d^2}{n^3 - n}$$

SIGNIFICANCE

As usual, the question of significance is, "How likely is it that the correlation observed in the sample is not a consequence of a correlation in the population that the sample was selected from but instead simply a result of the vagaries of sampling?" Put another way, "How likely is it that a sample this size with a correlation this strong could be selected from a population where there is no correlation?" For samples of 10 or more, this question can be answered with the familiar t table (Table 9.1). The following formula gives the value of t:

$$t = r_S \sqrt{\frac{n-2}{1-r_S^2}}$$

In our example,

$$t = .93 \sqrt{\frac{17-2}{1-.93^2}} = .93 \sqrt{\frac{15}{1-.86}} = .93 \sqrt{107.14} = 9.63$$

Looking this value up in Table 9.1, using the row for $n - 1 = 16$ degrees of freedom, we discover that this value of t would be far beyond the rightmost column in the table. The associated probability, then, would be far less than .001. Thus there is far less than 1 chance in 1,000 that a sample of 17 would show a Spearman's rank correlation this strong if it had been selected from a population where there was no rank order relationship between the two variables.

It should be noted that this example raises some complicated questions of what population the data are a sample from. The sample consists of 17 soil zones that have been surveyed. In order to accomplish the analysis we have just done, we must take these 17 soil zones as a random sample from a larger and vaguely defined population of soil zones that are or might be in the Konsankoro Plain. This sample has given us what we take to be 17 separate and independent observations for the two variables, and these 17 observations form the batch that we have analyzed as a sample. Strictly speaking, this is not a random sample from a population of soil zones. Indeed, this sample may represent a complete survey of the entire Konsankoro Plain. If

> ### Be Careful How You Say It
>
> In conclusion to the example analysis in the text, we would say "There is a strong and highly significant correlation between soil productivity and number of villages per square kilometer ($r_s = .93$, $p < .001$)." This informs the reader that the relationship is positive (more villages in more productive soil zones), what correlation coefficient was used, and just how unlikely it is that the observed correlation would have occurred in this sample if there were no correlation in the population from which the sample was selected.

we have studied the entire population, it may seem to make little sense to treat the data as a sample. In evaluating significance, however, we frequently engage in a sort of pretend sampling from an imaginary larger population. What we learn from the evaluation of significance in a case like this is still, however, whether we should have much confidence in the correlation observed. What we have found out in this instance is that the correlation we observed is not at all likely to be pure random chance at work in a small sample. We will consider this notion of pretend sampling further in the last chapter.

The formula for values of t given above is appropriate only if the sample is 10 or more. If the size of the sample is less than 10, then Table 15.2 should be used to determine the associated probability.

Table 15.2. Probability Values for Spearman's Rank Correlation (r_s) for Samples Less Than 10[a]

Confidence	80%	90%	95%	99%
	.80	.90	.95	.99
Significance	20%	10%	5%	1%
	.20	.10	.05	.01
n				
4	.639	.907	1.000	
5	.550	.734	.900	1.000
6	.449	.638	.829	.943
7	.390	.570	.714	.893
8	.352	.516	.643	.833
9	.324	.477	.600	.783

[a]Adapted from "Distributions of Sums of Squares of Rank Differences for Small Numbers of Individuals" by E. G. Olds [*Annals of Mathematical Statistics* 9:133–148 (1938)].

Statpacks

Spearman's rank correlation coefficient is only one of several similar approaches to evaluating the strength and significance of rank order correlations. Many statpacks provide options for calculating them all under the heading of rank correlations or nonparametric correlations. Sometimes, r_s is calculated as an option with the same commands that produce Pearson's r. Even if your statpack does not provide Spearman's rank correlation as a specific option, you still may be able to trick it into producing r_s. It turns out that Spearman's rank correlation is equivalent to Pearson's r calculated on rankings. Consequently, you can provide rankings for each of your cases on the variables you are interested in (the fourth and fifth columns in Table 15.1) and use your statpack to perform a regression analysis on those variables. The resulting correlation coefficient will be equivalent to r_s.

ASSUMPTIONS AND ROBUST METHODS

Since Spearman's rank correlation does not assume normal distributions, or rely on means, standard deviations, or scatter plots, it is automatically highly robust. No transformations or other modifications need ever be applied. This, in effect, makes r_s a very robust correlation coefficient that can be used instead of Pearson's r when such factors present problems for the application of Pearson's r.

PRACTICE

You have excavated the remains of 12 dwellings in the village site of Teixeira. You notice that some of the artifacts recovered from the dwelling areas are finer and fancier than others and might indicate differences in status or wealth between the households. You identify a variety of ornamental objects and pottery with incised decoration as possible status indicators, and you count the number of such artifacts in each household area per 100 artifacts recovered. This gives you an index of status or wealth based on the artifact assemblages in the different households. You wish to investigate whether this status index is related to the size of the dwelling structure itself (pursuing the idea that wealthier families might have larger houses). The data are given in Table 15.3.

1. How strong and how significant is the relationship between house floor area and your status index?

Table 15.3. Floor Area and Artifact Status Index for 12 Excavated Houses from the Teixeira Site

Status index	Floor area (m^2)	Status index	Floor area (m^2)
23.4	31.2	24.2	30.5
15.8	28.6	15.6	26.4
18.3	27.3	20.1	29.5
12.2	22.0	12.2	23.1
29.9	45.3	18.5	26.4
27.4	33.2	17.0	23.7

2. What sort of support do your observations provide for the idea that wealthier households (as indicated by their possessions) had larger houses?

Part IV
Special Topics in Sampling

Chapter *16*

Sampling a Population with Subgroups

When the population we are interested in has subgroups that we are also interested in separately, it is often useful to select a separate sample of elements from each of the subgroups. For such purposes each subgroup is treated as if it were a completely separate population. A sample of whatever size necessary or possible is selected from each of these separate populations, and the values of interest are estimated separately for each population. Suppose that we have reliable information on the locations of all sites in a region. No one has attempted to discover the sizes of these sites, however. We could select a sample of the known sites and go make systematic surface collections in an effort to determine how large they are. These determinations could then form the basis for estimating the mean site size for the region. If, in addition, the region could be divided into three different environmental settings (remnant levees, river bottoms, and slopes) we might be interested in estimating the mean site area for each of the settings.

Table 16.1 provides information on a sample of sites for each of these three settings, as well as a stem-and-leaf plot for each sample. The table gives N, the total number of sites in each setting (the three populations sampled), and n, the number of sites in each of the three samples. The stem-and-leaf plots show a single-peaked and symmetrical shape for each of the samples, and their standard errors have been calculated using the finite population corrector (Chapter 9) since the sampling fractions are large. Multiplying these standard errors by the corresponding values of t for 95% confidence and $n - 1$ degrees of freedom gives us error ranges to attach to the estimated mean site areas for each of the three settings. Thus we are 95% confident that

Table 16.1. Site Areas (ha) in Three Settings

River bottoms			Remnant levees			Slopes		
$N = 53$			$N = 76$			$N = 21$		
$n = 12$			$n = 19$			$n = 7$		
$\overline{X} = 2.78$			$\overline{X} = 1.71$			$\overline{X} = .83$		
$SE = .14$			$SE = .15$			$SE = .13$		
3.3			2.9			0.7		
2.7	4		1.7	4		1.3	4	
2.1	3	8	1.3	3		1.2	3	
3.8	3	134	2.1	3	2	0.6	3	
2.7	2	7789	1.9	2	59	0.6	2	
3.4	2	144	1.2	2	0113	1.2	2	
2.9	1	8	2.5	1	66779	0.2	1	
2.8	1		2.1	1	0234		1	223
2.4	0		1.6	0	78		0	667
1.8	0		1.7	0	4		0	2
2.4			2.0					
3.1			1.6					
			1.0					
			1.4					
			2.3					
			3.2					
			0.8					
			0.4					
			0.7					

the mean area of sites on remnant levees is 1.71 ha ± .32 ha; in the river bottoms, 2.78 ha ± .31 ha; and on the slopes, .83 ha ± .32 ha.

These estimates and their 95% confidence error ranges confirm what we might well have suspected from looking at the three stem-and-leaf plots—sites in the three settings have markedly different mean sizes, and the differences that we observe between our three samples are not at all likely to be just the result of sampling vagaries. Up to this point, we have done nothing more than treat these three samples in the ways discussed in Chapter 9.

POOLING ESTIMATES

At this point, however, we might well want to consider the three samples together in order to talk about sites in the region in general, irrespective of the settings in which they were located. We cannot simply put all the sites from all three samples together into one sample, though, and consider it a random sample of sites in the region. Such a sample would most definitely not be a random sample of the sites in the region because the selection procedures did not give each site in the region an equal chance of selection. Of the 21 sites on the slopes, 7 (or 33.3%) were selected; of the 53 sites in the river bottoms, 12 (or 22.6%) were selected; and of the 76 sites on remnant levees, 19 (or 25.0%) were selected. Thus river bottom sites had less chance of being included in the sample (a probability of .226) than sites on levees (a probability of .250), and levee sites had less chance of being included than sites on the slopes (a probability of .333). The overall sample produced by just putting these three separate samples together would systematically overrepresent slope sites and systematically underrepresent river bottom sites. Any conclusions we might arrive at about mean site area in the region as a whole based on such a sample would be affected by these sampling biases.

What we must do is consider the larger problem one of *stratified sampling*, as selecting separate samples from different subgroups of a population is usually called. In this example, each of the three environmental settings would be a *sampling stratum*. Each sampling stratum would form a population to be sampled separately from the other sampling strata, just as we have done in this example. Appropriate sample sizes and sampling procedures would be determined independently for each sampling stratum, and the samples selected would be used independently to make estimates about each of the parent populations. We have already done all of this. It raises no new issues in sampling beyond those dealt with in Chapters 7–10.

Only at the last step, that of *pooling* the estimates made for each sampling stratum into an overall estimate for the whole population, must special steps be taken. In the first place, having already discovered that sites in the three different settings have rather different mean areas, we must consider whether it makes any sense even to speak of the mean area of sites for the region as a whole. If the overall population of sites had a shape with multiple peaks, it would be foolish to attempt any analysis of the entire set of sites as a single batch. We do not, of course, have any way of knowing for certain what the shape of the whole population would be, but, since the sampling fractions in the three sampling strata are not wildly different, we could look at a stem-and-leaf plot of all three samples together to get a rough idea. Such a stem-and-leaf plot appears in Table 16.2. It is certainly single-peaked and symmetrical enough to make it meaningful to use the mean as an index of center for the whole batch. Thus, we could consider it sensible to make an estimate of the mean site area for all sites in the region by pooling the estimates for the three sampling strata, as follows:

$$\overline{X}_p = \frac{\sum (N_h \overline{X}_h)}{N}$$

**Table 16.2. Stem-and-Leaf Plot
of Areas of Sites from All
Three Samples in Table 16.1**

4	
3	8
3	1234
2	577899
2	0111344
1	667789
1	0222334
0	66778
0	24

where

$\bar{X}_p$ = the pooled estimate of the mean, that is, the estimated mean for the
entire population, taking all sampling strata together;
$\bar{X}_h$ = the mean of the elements in the sample for stratum h;
N_h = the total number of elements in the population of stratum h;
N = the total number of elements in the entire population.

For the example from Table 16.1,

$$\bar{X}_p = \frac{(76)(1.71) + (53)(2.78) + (21)(.83)}{150} = \frac{294.73}{150} = 1.96 \, \text{ha}$$

Thus we estimate that the mean area of sites in the region as a whole (irrespective of environmental setting) is 1.96 ha. We attach an error range to this estimate in a similar fashion, by pooling the standard errors for the three separately selected samples:

$$SE_p = \frac{\sqrt{\sum(N_h^2)(SE_h^2)}}{N}$$

where

SE_p = the pooled standard error for all sampling strata taken together;
SE_h = the standard error for sampling stratum h;
N_h = the total number of elements in the population of stratum h (as before);
N = the total number of elements in the entire population (also as before).

For the example from Table 16.1,

$$SE_p = \frac{\sqrt{(76^2)(.15^2) + (53^2)(.14^2) + (21^2)(.13^2)}}{150} = \frac{13.87}{150} = .09$$

This pooled standard error is treated like any other. To produce an error range for 95% confidence, we would multiply it by the value of t corresponding to 95% confidence and $n-1$ degrees of freedom, where n is now the number in all three samples considered together, or 38. This value of t is 2.021, so we would be 95% confident that the mean area of all sites in the region is 1.96 ha $\pm$ .18 ha.

THE BENEFITS OF STRATIFIED SAMPLING

Stratified sampling can sometimes offer a more precise estimate for an entire population than simply sampling the entire population directly. This makes stratified sampling potentially useful even in situations where we might not be much interested in the separate means of the sampling strata. The possible increased precision comes from providing a smaller error range in the situation where a population has subgroups whose means differ somewhat from each other but which have very small standard deviations when each is taken separately. That is, if the subgroups each form batches with smaller spreads than the population as a whole, the error ranges associated with the estimates of their means may be quite small. When these are pooled into an error range for the estimated overall population mean it may well be smaller than the error range that would have been obtained from a single sample drawn randomly from the population as a whole. Sometimes this effect is strong enough to outweigh the opposite effect resulting from the fact that the samples from the subgroups are each smaller than the total sample. If a population is easily divided into subgroups whose means may be different and whose members vary little from each other, then, it is worth considering sampling that population by those subgroups instead of as a whole, even if the subgroups are of little intrinsic interest separately.

Chapter *17*

Sampling a Site or Region with Spatial Units

Sometimes the sampling elements available for selection are not the same as the elements we wish to study. This happens most frequently in archaeology in *spatially based sampling,* as in the excavation of a sample of grid squares in a site or the survey of a sample of grid squares or transects in a region. For instance, suppose we have a random sample of 500 sherds from a site. We may want to estimate, say, the mean thickness of sherds at the site or the percentage of a particular pottery type in the sherds at the site. The elements studied are sherds. Suppose the sample had been obtained by excavating a random sample of 10 grid squares. The sampling element here is not the sherd but the grid square. It was 10 grid squares that were randomly selected from all the squares in the site grid, not 500 sherds from all the sherds in the site. We thus have a sample, not of 500 independently selected elements, but of 10 independently selected elements, and these elements do not correspond to the elements we need to study. Each sampling element is, in this case, a group or *cluster* of a varying number of the elements of study (sherds). This fact must be allowed for in making estimates of means or proportions.

Estimating population means and proportions from samples and attaching error ranges to those estimates was the subject of Chapters 9 and 10. This chapter extends that discussion to the special case where the sampling elements are different from the elements of study. This chapter on *cluster sampling,* then, can be considered a special case of the general topics dealt with in Chapters 9 and 10, which can be referred to as *simple random sam-*

pling to distinguish them from more complex kinds of sampling. Cluster sampling is particularly important in archaeology because so much of the sampling we do is based on spatial units.

SELECTING A SAMPLE OF SPATIAL UNITS

At least three different kinds of spatial sampling units might be used in archaeology—*points, transects,* and *quadrats*—but points and transects are very rarely used. *Quadrats* (not *quadrants,* which are something else) are two-dimensional spatial units, in archaeology most often the squares in a grid system. They can also be rectangles or other shapes. When the rectangles are long and narrow and run from one side of the study area to the other, we often refer to them as *transects,* but technically these are quadrats. True transects, like lines, have only length; their width is 0. When an archaeologist walks along a "transect" from one side of a survey zone to the other, the observations are not actually along a line but within a very long narrow rectangle including some distance to either side of the path walked. Such "transects" are usually best treated as long, narrow quadrats in cluster sampling since they do have a width (and thus an area) based on how far to either side the archaeologist can observe whatever is to be observed.

Perhaps the most frequently used method of selecting a random sample of quadrats is to lay out a grid dividing the area to be sampled (say, a site to be excavated) into sampling units. Each potential excavation unit in the grid system can be assigned a number beginning with 1, and a random number table can be used to select a sample of these quadrats. One possible result of

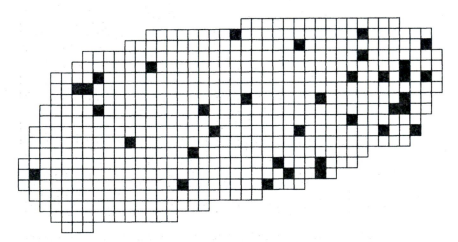

Figure 17.1. A random sample of quadrats selected individually.

such a sampling scheme is shown in Figure 17.1. The same system can be used for long, narrow quadrats ("transects"). In this case the grid divides the area to be sampled into long, narrow rectangles running from one side to the other, each as wide as the coverage of a single "transect." These are assigned numbers for random selection (Figure 17.2).

Another interesting possibility, sometimes used to avoid having all "transects" parallel to each other, is to enclose the area to be sampled in a rectangular frame on a map and place tick marks on all sides of the frame. The ticks should be as far apart as the "transects" are wide. The ticks are numbered sequentially, beginning at any point with 1 and continuing all the way around the frame until the starting point is reached again. A random number determines one end of the first "transect," and a second random number determines its other end. (If the second random number indicates a tick mark on the same side of the frame as the first, it is discarded and another is selected.) The process is repeated until the desired number of "transects" has been selected. One result of such a sampling scheme is illustrated in Figure 17.3.

When n random quadrats are selected from all the quadrats in the grid at large, it is often the case that some sample quadrats are very close to each other (possibly even adjacent), and one or more fairly large parts of the study area may be left entirely unsampled. Figure 17.1 shows both of these characteristics. This may be unsatisfactory, for example, in excavating a site by random sampling, since (for reasons not related to random sampling) we might not want to leave one whole section untested. One alternative sometimes applied in such situations is *systematic sampling*. As an example, sup-

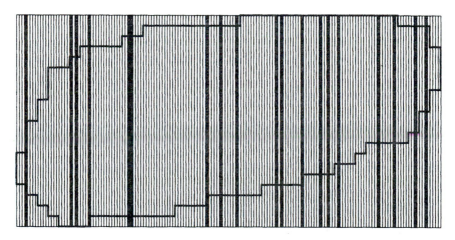

Figure 17.2. A random sample of "transects" (actually very long, narrow quadrats) selected individually.

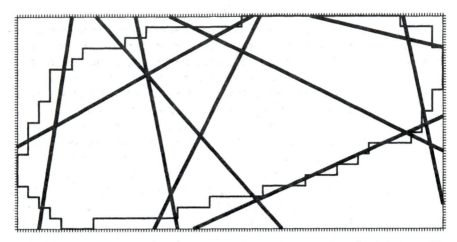

Figure 17.3. A random sample of "transects" (actually very long, narrow quadrats) determined by random selection of their endpoints.

pose that we want a sample of 35 quadrats from an area consisting of 570 quadrats. To select a systematic sample, we would subdivide the grid of 570 quadrats into 36 subsets consisting of 16 contiguous quadrats each. (We would add six dummy quadrats, indicated with dots in Figure 17.4, to fill out the full 16 in each subset.) One quadrat would be randomly selected from each subset of 16 by repeatedly selecting random numbers between 1 and 16. The sample might be as large as 36, but if any dummy quadrats are selected,

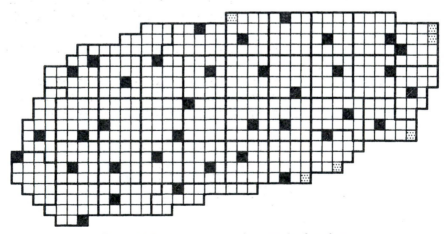

Figure 17.4. A systematic random sample of quadrats.

then the final sample is less than 36. The dummy quadrats never really become part of the sample, even if they are selected, because they are imaginary. Their function is to provide each real quadrat exactly 1 chance in 16 of being selected for the sample. The resulting sample could still include adjacent quadrats, as Figure 17.4 shows, but the large unsampled areas that frequently occur in simple random samples of quadrats would be impossible.

It is sometimes objected that systematic sampling is not strictly random, and technically this is true. In strict terms, the selection of one element in a random sample should not have any effect at all on the selection of other elements. The selection of a quadrat in a systematic sample, however, causes the other quadrats in the same subset to lose their eligibility for future selection. Perhaps more important, systematic sample selection, as described here, comprises sampling without replacement since a quadrat, once selected, is no longer available for future selection. The equations given in this chapter, like those in Chapters 9 and 10, are for sampling with replacement. The impact of these two technical problems, however, is minimal under most circumstances. (And especially for the form of systematic sampling described here, which violates the strictest norms of random sampling less than other variants of systematic sampling that have been suggested.) The attractiveness of working with a spatial sample that avoids leaving any large sections of the study area unexamined usually outweighs these minor technical objections.

ESTIMATING POPULATION PROPORTIONS

The best estimate of a population proportion in cluster sampling is the same as in the case of simple random sampling—simply the proportion in the sample. For instance, suppose we excavate a random sample of 10 grid squares in a site and obtain 500 sherds altogether. If 35% of this cluster sample of sherds are cord-marked, we would estimate that 35% of the sherds in the site are cord-marked.

The error range corresponding to this estimate, as with simple random sampling, is based on the standard error of the proportion, but the standard error of the proportion in cluster sampling is calculated by the formula

$$SE = \sqrt{\left(\frac{1}{n}\right)\left(\frac{\sum\left(\frac{x}{y} - P\right)^2 \left(\frac{yn}{Y}\right)^2}{n-1}\right)\left(1 - \frac{n}{N}\right)}$$

where

SE = standard error of the proportion;
n = sample size (i.e., number of units in the sample);
N = population size (i.e., number of units in the population);
x = number of object x in a unit;
y = number of object y in a unit;
P = estimate of the proportion x/y for the population;
Y = Σy, that is, the total number of object y in n units.

Note that this formula incorporates the finite population corrector $1 - (n/N)$ discussed in Chapter 9. Since the population is usually finite and definable in spatial sampling, this correction can usually be applied. If the population is very large, however, compared to the size of the sample, the finite population corrector has a negligible effect on the result because it is only trivially different from one.

Table 17.1 provides an example of the calculations involved. It describes a random sample of 10 excavated grid units from a site whose total area is 100 grid units. The sample size, n, is thus 10, and the population size, N, is 100. These 10 excavation units yielded a total of 500 sherds, some of which were cord-marked. We wish to estimate the proportion of cord-marked sherds in the ceramic assemblage of the site as a whole. The units are identified by their sequential numbers, which were used to select the random sample. Since the proportion we wish to estimate is the proportion of cord-marked sherds in the ceramic assemblage, x is the number of cord-marked sherds in each square and y is the total number of sherds in each square. Σx, or X, is 175—the total number of cord-marked sherds found in all 10 units. And Σy, or Y, is 500—the total number of sherds found in all 10 units. The

Table 17.1. Sherds from a Random Sample of 10 Excavation Units

Unit	x No. of cord-marked sherds	y No. of sherds
07	10	32
18	13	27
29	16	38
31	19	73
37	17	55
56	21	41
72	18	63
83	30	81
87	19	56
91	12	34

proportion of cord-marked sherds in the ceramic assemblage for the entire sample, then, is $(175/500) = .350$; 35.0% of the sherds are cord-marked. Since the best estimate of the population proportion is the sample proportion, we would estimate that 35.0% of the sherds in the site are cord-marked.

Table 17.2 extends Table 17.1 into a step-by-step calculation of the summation needed for the standard error calculation. The first calculation step (x/y) results in the fourth column, simply dividing x by y for each excavation unit. This quantity is, of course, the proportion of cord-marked sherds for each of the 10 excavated units. This proportion varies from a low of 26.0% in unit 31 to a high of 51.2% in unit 56. It is this variation from one sample unit to the next in the proportion of interest to us that will form the basis for the error range.

For the fifth column in the table, $x/y - P$, the overall sample proportion $(.350)$ is subtracted from each excavation unit's proportion. This is recognizable, of course, as a deviation—the extent to which the proportion of cord-marked sherds in each excavation unit deviates from the overall proportion in the whole sample taken together. In a manner familiar from all our calculations of standard deviations, the next step, $(x/y - P)^2$, squares the deviations from the fifth column to arrive at the sixth column. The sixth column is one of the two terms that must be multiplied together to arrive at the quantity to be summed.

The other term is, in effect, a weighting factor. The sixth column is a set of squared deviations. In cluster sampling we weigh more heavily the deviations of units that produce more evidence (that is, in this example,

Table 17.2. Calculation of the Summed Weighted Deviations from the Overall Sample Proportion

Unit	x	y	$\dfrac{x}{y}$	$\dfrac{x}{y} - P$	$\left(\dfrac{x}{y} - P\right)^2$	$\dfrac{yn}{Y}$	$\left(\dfrac{yn}{Y}\right)^2$	$\left(\dfrac{x}{y} - P\right)^2\left(\dfrac{yn}{Y}\right)^2$
07	10	32	.313	−.037	.001369	.640	.409600	.000561
18	13	27	.481	.131	.017161	.540	.291600	.005004
29	16	38	.421	.071	.005041	.760	.577600	.002912
31	19	73	.260	−.090	.008100	1.460	2.131600	.017265
37	17	55	.309	−.041	.001681	1.100	1.210000	.002034
56	21	41	.512	.162	.026244	.820	0.672400	.017646
72	18	63	.286	−.064	.004096	1.260	1.587600	.006503
83	30	81	.370	.020	.000400	1.620	2.624400	.001050
87	19	56	.339	−.011	.000121	1.120	1.254400	.000152
91	12	34	.353	.003	.000009	.680	.462400	.000004

$$\Sigma y = 500 \qquad\qquad \Sigma\left(\frac{x}{y} - P\right)^2\left(\frac{yn}{Y}\right)^2 = .053132$$

more sherds). This makes some intuitive good sense if you remember that generally we get a more accurate estimate from a larger sample than from a smaller sample. In effect, each excavation unit is a single sample of sherds from the site. These single samples can be expected to produce somewhat different results—that is, to deviate from the overall proportion. If they all deviate very little from the overall proportion, then the error range associated with our cluster sampling estimate should be relatively small. (The consistency from one excavation unit to another makes us willing to believe the site is fairly homogeneous and our estimate fairly precise.) We will be more concerned about imprecision in our results if units that produce large numbers of sherds deviate widely from the overall proportion rather than if units producing small numbers of sherds deviate widely from the overall proportion. Thus, when we sum up the squared deviations, we will count the deviations in units that produced large numbers of sherds more heavily than deviations in units that produced small numbers of sherds (where the deviations are more likely to be simply the result of random vagaries in smaller samples).

The seventh column begins the calculation of this weighting factor, which is based on how "large" the sample unit is in terms of the elements we are studying. In this example the elements we are studying are sherds, so the large (heavily weighted) sample units are those that produced large numbers of sherds. The seventh column (yn/Y) is simply the number of sherds in a unit times the number of units in the sample (10) divided by the total number of sherds in the sample (500). (It is useful to know, for checking calculations, that the sum of the seventh column is always n, the number of units in the sample.) The eighth column is simply the square of the seventh column.

The last column in Table 17.2 is the product of the fifth column and the eighth column and is the quantity to be summed for all 10 sample units: $(x/y - P)^2 (yn/Y)^2$. The sum of this quantity for all sample units is indicated at the bottom of the last column: .053132.

Substitution of numbers into the formula given above for the standard error of the proportion is now relatively straightforward:

$$SE = \sqrt{\left(\frac{1}{10}\right)\left(\frac{.053132}{10-1}\right)\left(1 - \frac{10}{100}\right)}$$

$$= \sqrt{(.100)(.005904)(.900)}$$

$$= \sqrt{.00531} = .023$$

Statpacks

Very few computer statpacks provide for the calculation of standard errors for cluster samples. They can certainly be calculated by hand, although the summation that forms the numerator of the middle fraction is tedious. The ease with which this calculation is illustrated as a table in which each column is derived by relatively simple repeated calculation from the previous one, however, suggests a computerized solution. Spreadsheet programs were designed precisely for performing such calculations, and are probably the fastest, easiest, and most commonly available option for getting a computer to do most of the boring work. The provisions that statpacks provide for transforming variables can also often be adapted to this task, since each column in Tables 17.2 and 17.4 really is simply a new variable whose values are calculated by a repeated mathematical manipulation from previous columns. Some database managers also provide mathematical tools that can perform calculations like this.

The standard error of the proportion, then, is .023, so the estimate of the proportion of cord-marked sherds in the ceramic assemblage at the site could take the form 35.0% ±2.3%. This error range can be increased by multiplying it by an appropriate value of t to make it a statement at whatever level of confidence is desired (see Chapter 9). For example, if we wish to express our estimate with an error range at the 95% confidence level, we look in Table 9.1 for the value of t corresponding to 95% confidence and 9 degrees of freedom $(n-1)$. This value is 2.262. Thus the error range we seek is $(2.262)(.023) = .052$. We would thus say that we are 95% confident that cord-marked sherds represent 35.0% ± 5.2% of the sherds at this site.

ESTIMATING POPULATION MEANS

As with proportions, the best estimate of the population mean is the overall mean in the sample. Table 17.3 provides example data from the same sample of excavated units we have been considering. The data here are lengths in millimeters of the projectile points encountered in these same excavation units. Altogether 21 projectile points were recovered, with an overall mean length of 21.8 mm. We would thus estimate that the mean length of all the projectile points in the site is 21.8 mm.

The standard error we need in order to put an error range with this estimated mean is calculated in a manner very similar to the standard error of the proportion:

Table 17.3. Lengths of Projectile Points from a Sample of 10 Excavation Units

Unit	x Projectile point lengths (mm)
07	15, 19, 23
18	17
29	18, 23
31	18, 18, 27
37	18, 19
56	24
72	20, 21, 26, 28, 29
83	16
87	28
91	25, 26

$$SE = \sqrt{\left(\frac{1}{n}\right)\left(\frac{\sum(\bar{x}-\bar{X})^2\left(\frac{yn}{Y}\right)^2}{n-1}\right)\left(1-\frac{n}{N}\right)}$$

where

SE = standard error of the mean;
n = sample size (i.e., number of clusters in the sample);
N = population size (i.e., number of clusters in the population);
$\bar{x}$ = mean of x in a cluster;
$\bar{X}$ = estimated population mean of x (i.e., the overall sample mean);
y = number of things in a cluster measured for x;
Y = Σy (i.e., the total of y in n clusters).

Once again, the complicated calculation is the summation that forms the numerator of the middle fraction, and this summation is quite similar to the summation required for the standard error of the proportion. Table 17.4 shows this calculation carried out.

In this instance, y is the number of projectile points found in each unit. For each excavation unit, we calculate a mean projectile point length based on the projectile points in that unit. This mean appears in the third column of Table 17.4. The deviation we are interested in this time is the difference

Table 17.4. Calculation of Summed Deviations from Overall Sample Mean

Unit	y	$\bar{x}$	$\bar{x}-\bar{X}$	$(\bar{x}-\bar{X})^2$	$\dfrac{yn}{Y}$	$\left(\dfrac{yn}{Y}\right)^2$	$(\bar{x}-\bar{X})^2\left(\dfrac{yn}{Y}\right)^2$
07	3	19.0	−2.8	7.840	1.429	2.042	16.009
18	1	17.0	−4.8	23.040	.476	.227	5.230
29	2	20.5	−1.3	1.690	.952	.906	1.531
31	3	21.0	−0.8	.640	1.429	2.042	1.307
37	2	18.5	−3.3	10.890	.952	.906	9.866
56	1	24.0	2.2	4.840	.476	.227	1.099
72	5	24.8	3.0	9.000	2.380	5.664	50.976
83	1	16.0	−5.8	33.640	.476	.227	7.636
87	1	28.0	6.2	38.440	.476	.227	8.726
91	2	25.5	3.7	13.690	.952	.906	12.403

$\bar{X} = 21.8$

$$\sum (\bar{x}-\bar{X})^2\left(\frac{yn}{Y}\right)^2 = 114.783$$

between the mean projectile point length for each unit and the overall mean in the sample (the fourth column). As usual, this deviation is squared (the fifth column). The weighting factor works just as it did in the case of estimating proportions (the sixth and seventh columns). The final product is summed in the last column.

The results can then be substituted in the formula on the preceding page as follows:

$$SE = \sqrt{\left(\frac{1}{10}\right)\left(\frac{114.783}{10-1}\right)\left(1-\frac{10}{100}\right)}$$

$$= \sqrt{(.100)\,(12.7537)\,(.900)}$$

$$= \sqrt{1.1478} = 1.07 \text{ mm}$$

To use this standard error as an error range at the 95% confidence level, as we did with the standard error of the proportion earlier in the example, we would multiply it by the value of t for 9 degrees of freedom and 95% confidence, which continues to be 2.262. Thus the error range would be $(1.07)(2.262) = 2.42$ mm, and we would be 95% confident that the mean length of all the projectile points in the site is 21.8 mm ± 2.4 mm.

DENSITIES

The fact that spatially based sampling often takes us into the realm of cluster sampling rather than simple random sampling should not confuse us about what the basic principle of cluster sampling is. Sometimes spatially based sampling is actually simple random sampling. It depends on what the elements to be studied are. In the two examples discussed above the elements to be studied were sherds and projectile points. Neither of these elements was the same as the sampling unit, since quadrats from a grid were the things randomly chosen to define the sample. Thus it was necessary to consider both estimating proportions of cord-marked sherds and estimating mean projectile point length as questions in cluster sampling.

Often, however, we are interested in studying the very spatial units that are randomly selected to form the sample. This happens, for instance, when we estimate the density (the number per unit area) of some artifact or feature. Such densities are usually easily expressed as numbers of things, say sherds, per grid unit. Such numbers are actually properties, not of the sherds, but rather of the grid units. The elements being studied are the same as the elements that were randomly selected to form the sample, so this becomes a question in simple random sampling. In the example above, we had a sample of 10 grid units that produced, respectively, 32, 27, 38, 73, 55, 41, 63, 81, 56, and 34 sherds. We can estimate the mean number of sherds per grid unit in the complete population of grid units (that is, the entire area of the site) in exactly the manner discussed in Chapter 9 for a sample of 10 with a measurement for each of the 10 units.

The mean number of sherds per grid unit in the sample is 50.0. Thus we would estimate that in the site as a whole the mean number of sherds per grid unit is 50.0. The standard error is 5.8 sherds, so an error range at the 95% confidence level would be $(2.262)(5.8) = 13.1$ sherds. We are thus 95% confident that the mean number of sherds per grid unit in the site is 50.0 ± 13.1 sherds.

Such estimates of densities are the most direct springboard to estimates of the total quantities of various things in the site. For example, we have estimated that there is an average of 50.0 ± 13.1 sherds per grid unit in the site (at a 95% confidence level). We know that the entire site consists of 100 grid units. Our density estimate thus translates into an estimate that the total number of sherds in the site is (50.0 *sherds per grid unit*) (100 *grid units*) = 5,000 *sherds*. The error range also translates in the same way: (13.1 *sherds per grid unit*) (100 *grid units*) = 1,310 sherds. We are thus 95% confident that the total number of sherds in the site is 5,000 ± 1,310.

Chapter *18*

Sampling without Finding Anything

Sampling statistics ordinarily take as their point of departure some finding in a sample. Say the sample consists of artifacts, including some projectile points. We can estimate the proportion of projectile points in the artifact assemblage from which the sample came; we can estimate the mean weight of projectile points for the population of projectile points from which the sample came; we can estimate the proportions of different raw materials of which projectile points in the population were made; and so on. Following the procedures discussed in Chapters 9, 10, and 17, we can attach error ranges for particular confidence levels to these estimates.

Sometimes, however, we have particular reason to be interested in some specific category of observation that just does not appear at all in a sample. For example, we may recognize chert, flint, and obsidian as potentially available raw materials from which projectile points could be made, but perhaps our sample includes only chert and flint points. How confidently can we say that obsidian was not used to make projectile points? We certainly know enough about samples by now to know that the fact that we find no obsidian projectile points in a sample does not necessarily mean that there were none at all in the population from which the sample was drawn. This is true no matter how large the sample is. The only way to be certain that there are no obsidian projectile points is to acquire and study the entire population of projectile points. As long as one projectile point remains unexamined, there is at least some possibility that it could be made of obsidian.

As long as we are working with a sample, then, we must settle for some level of confidence short of 100%, just as in all the conclusions we have made about populations on the basis of samples. The confidence we have, as always, will depend on the size of the sample. We will be more confident in saying that projectile points were not made of obsidian if we have failed to find any obsidian points in a sample of 100 points than if we have failed to find them in a sample of 5 points. The proportion of projectile points in the population is also involved. It is intuitively obvious that if the population really includes many obsidian projectile points, it is more likely that at least one will turn up in a sample of a given size than if there are very few obsidian points in the population.

The aim of this chapter is to put a finer point on these intuitive (but perfectly valid) approximations. Applying basic statistical principles to the task requires only deciding at what level of confidence we need to speak and how many obsidian projectile points we are willing to risk overlooking. Deciding at what level of confidence to speak presents no novel aspect in this context; all of the considerations brought to bear in previous chapters apply. Deciding how many projectile points we are willing to risk overlooking, however, does raise a new issue, and making this decision is what enables us to put statistical tools to good use here. In effect, we must decide what low proportion of obsidian projectile points is functionally equivalent to none. If only one projectile point in a billion were made of obsidian, we would presumably be willing (for many purposes at least) to say that, in effect, obsidian was not used for projectile points. We would probably be equally willing to say that one obsidian point in a million really meant that points were not made of obsidian. For some purposes at least, it would be interesting and useful to be able to say with high confidence that fewer than 1% or even 5% of the projectile points in some population were made of obsidian.

Suppose that we have a sample of 16 projectile points, and none is made of obsidian. We would like to know at what level of confidence we can say that fewer than 1% of the projectile points in the population from which the sample was selected were made of obsidian. Another (and more familiar) way to put this question is, "How likely is it that we could select a random sample of 16 including no obsidian projectile points from a population with as many as 1% obsidian projectile points?" Answering this question is simply a matter of multiplying the probabilities of a series of sequential events.

Assume that 1% of the population of projectile points we have sampled actually are made of obsidian. The probability that the first point we select for our sample will *not* be made of obsidian is .99. (Since 99% are not made of obsidian, 99 times out of 100 a randomly selected point will not be made of obsidian.) There is also a probability of 99% that the second point selected for the sample will not be made of obsidian. Thus, 99% of the time we will select a non-obsidian point first. If this happens, then 99% of the time we will select a non-obsidian point second. Thus 99% of 99% of the samples of two from this population will not include obsidian points. The probability of drawing a sample of two with no obsidian points from a population with 1% obsidian points, then, is (.99) (.99) = .980, or 98%.

If we repeatedly select samples of two from this population, then, 98.0% of those samples will contain no obsidian points. Having found no obsidian points in a sample of two, we might continue to enlarge the sample by selecting a third random point. This third random projectile point, like any randomly selected point, will not be made of obsidian 99% of the time. Thus, in repeatedly drawing samples from this population, 98.0% of the time we will not find an obsidian point among the first two selected, and in 99% of those 98.0% of the instances, when we continue to select a third point we will still not have found one made of obsidian. Thus there is a probability of $(.99)(.980) = .970$ that a sample of three points from this population will not contain an obsidian point. We can continue in this fashion to select more and more points. At each step the probability from the previous step is multiplied once again by .99.

For any sample size, n, then, the probability of selecting a sample with no obsidian points from this population is $.99^n$. Thus, for a population with as many as 1% obsidian projectile points the probability that a sample of 16 would contain no obsidian points is $.99^{16} = .851$. For a sample of 50 points, there would still be a 60.5% chance of selecting a sample with no obsidian points from a population with 1% obsidian points ($.99^{50} = .605$). This probability of 60.5% amounts to an evaluation of significance. That is, it is the probability that a random sample of 50 may contain no obsidian points even though the population does have as many as 1% obsidian points. The opposite probability ($1 - .605 = .395$) is the confidence level at which we could say that the population from which our sample was selected has fewer than 1% obsidian points. Thus a sample of 50 with no obsidian points would give us only 39.5% confidence that it was selected from a population with fewer than 1% obsidian points.

Table 18.1 provides the confidence levels for given sample sizes (n) and given population proportions. The figures for the example just discussed can be found there by looking at the row for $n = 50$ and the column for a population proportion of 1%. The number in the table is .395, corresponding to a 39.5% level of confidence that the population from which a sample of 50 elements was drawn actually has fewer than 1% of whatever item of interest it was that failed to appear in the sample. A confidence level of 39.5% is, of course, not a very useful level of confidence at which to speak. To determine how large a sample of projectile points without any made of obsidian we would need in order to make this conclusion at the 95% confidence level, we can read farther down the column for a population proportion of 1% until we reach .95. In the row corresponding to a sample size of 300, the confidence level has finally reached .951. Thus, if we wanted a sample large enough to conclude at the 95% confidence level that the population had fewer than 1% obsidian points, we would need a sample of some 300 projectile points. (If we did find one or more obsidian projectile points in this enlarged sample, of

Table 18.1. Confidence Levels for Concluding That Absence from a Sample Indicates a Low Population Proportion

Population proportion n	0.1%	0.5%	1.0%	2.0%	5.0%
20	.020	.095	.182	.332	.642
25	.025	.118	.222	.397	.723
30	.030	.140	.260	.455	.785
35	.034	.161	.297	.507	.834
40	.039	.182	.331	.554	.871
45	.044	.202	.364	.597	.901
50	.049	.222	.395	.636	.923
55	.054	.241	.425	.671	.940
60	.058	.260	.453	.702	.954
70	.068	.296	.505	.757	.972
80	.077	.330	.552	.801	.983
90	.086	.363	.595	.838	.990
100	.095	.394	.634	.867	.994
110	.104	.424	.669	.892	.996
120	.113	.452	.701	.911	.998
130	.122	.479	.729	.928	.999
150	.139	.529	.779	.952	>.999
175	.161	.584	.828	.971	—
200	.181	.633	.866	.982	—
250	.221	.714	.919	.994	—
300	.259	.778	.951	.998	—
350	.295	.827	.970	.999	—
400	.330	.865	.982	>.999	—
450	.363	.895	.989	—	—
500	.394	.918	.993	—	—
600	.451	.951	.998	—	—
700	.504	.970	.999	—	—
800	.551	.982	>.999	—	—
900	.594	.989	—	—	—
1000	.632	.993	—	—	—
1200	.699	.998	—	—	—
1400	.754	.999	—	—	—
1600	.798	>.999	—	—	—
1800	.835	—	—	—	—
2000	.865	—	—	—	—
2500	.918	—	—	—	—
3000	.950	—	—	—	—
4000	.982	—	—	—	—
5000	.993	—	—	—	—
6000	.998	—	—	—	—
7000	.999	—	—	—	—

course, we would simply turn back to the procedures discussed in Chapter 10 to estimate the proportion in the population and attach an error range at the 95% confidence level to this estimate.)

Table 18.1, then, can be used to determine the level of confidence at which we can conclude that something absent from a sample of a given size occurs in the population in a proportion of less than 5%, 2%, 1%, 0.5%, or 0.1%. It can also be used to determine how large a sample will be needed to conclude at a given confidence level that the population contains less than a certain proportion of some item of interest. As can readily be seen, if we need high confidence that the population proportion for an item absent from the sample is very low, then quite a large sample is required.

Chapter *19*

Sampling and Reality

At the beginning of Chapter 7, I asserted that sampling was at the very heart of the statistical principles applied in this book. I hope the chapters that lie between there and here have made clearer just what that means. Whether the task is estimating the population mean or proportion, comparing means in several batches, comparing proportions in several batches, or investigating the relationship between two measurements, the logic of the approaches statisticians take involves thinking about the batches of numbers we are working with as samples from a larger population. It is this larger population that really interests us.

Sometimes this is literally and obviously true. If, for example, we excavate an entire rock shelter site and recover 452,516 pieces of lithic debitage, we might well select some kind of random sample of this debitage for detailed analysis with the objective of characterizing the entire population of 452,516 waste flakes. In this case we would have a sample of waste flakes that we would use to make statements about the population of all debitage at the site from which the sample was selected. The sampling design we used might well be rather complicated. For instance, we might want to be able to compare one stratum in the site to the others, so we might separately select a sample from each stratum. The techniques discussed in Chapters 9 and 10 would enable us to determine approximately how large each of these samples would need to be in order to accomplish our aims, and they would enable us to estimate means of measurements we might make and the proportions of different categories we might define in the several populations consisting of debitage from each stratum. We could attach error ranges to these estimates that would help us to know at what confidence level and with what precision we could discuss these means and proportions (Chapters 9 and 10). We could compare the means of the measurements in different strata using these estimates and error ranges or using t tests and analysis of variance (Chapters 11 and 12). We could compare the proportions of the categories in different strata using the estimates and error ranges or using chi-square (Chapter 13).

We could evaluate the strength and significance of relationships between measurements with a regression analysis (Chapter 14). If we had ranks rather than true measurements, we could use a rank correlation (Chapter 15). We could combine samples from different strata to say things about the debitage from the site as a whole (Chapter 16). If there were some category of material that just did not appear in our sample, we could evaluate the confidence with which we could talk about its rarity in the population from which the sample came (Chapter 18). All these analyses would provide us with ways to say how much confidence we had that some observation of interest in the sample reflected something that occurred in the population from which the sample was selected as well. This is a very straightforward application of the principles in Chapters 9 through 18.

If we had a large sample of Early, Middle, and Late Archaic projectile points, we might well do all of these same things with it—means and proportions with error ranges, significance tests of the differences between Early and Middle or Middle and Late, and so on. We have to stop and think for a minute, though, about just what population it is that we are talking about on the basis of our sample. Probably, the population is something like the large and very vaguely defined set of all projectile points made during the Early, Middle, and Late Archaic. Our interest is likely to be in identifying some kind of change from one period to the next in very general terms. We still clearly have a sample, and we can imagine the population we are talking about even if it is a fairly nebulous population. The sample has not been selected with truly random procedures, so the issue of sampling bias is highly relevant in this example, unlike the previous one (Chapter 7).

If we excavated a Formative village in its entirety, recovering information about 27 house structures from Early, Middle, and Late Formative, the same list of statistical tools might be put to use. If we had excavated all the houses at the site, though, it is even less clear what sense it makes to talk about these houses as a sample. What is the population they are a sample from? Aren't they the complete population? And does this mean we can't investigate the significance of a difference we might observe between, say, Early and Middle Formative? In a case like this, there are several kinds of populations we might implicitly be interested in. One is the population of all houses that existed at the site at any point in the Formative. Some have surely been destroyed by the construction of subsequent houses and other processes. Our sample is not this complete population, but in some contexts this would be the complete population of interest to us. In other contexts, we might use the sample of excavated Early Formative houses from this site as a way of talking about Early Formative houses in general. The relation between our sample and the population of interest in this context is similar to the first example concerning Archaic projectile points.

If we surveyed a whole region completely, with 100% coverage, it would become even more difficult to identify just what population we take our sample to represent. Presumably some sites of both periods would have been destroyed or made inaccessible to survey, but if conditions were so propitious for survey that we recovered data on almost all the sites, talking about our sites as a sample from a larger population has become very forced. What would it mean to talk about, say, the significance of a difference in mean site size between the Neolithic and the Bronze Age? We could certainly perform the calculations necessary to say that the mean site area in the Neolithic was 1.4 ± .2 ha at the 95% confidence level and in the Bronze Age it had changed to 3.6 ± .3 ha. (Or, instead, we could perform a *t* test and find out that the significance of the difference in mean site area between Neolithic and Bronze Age sites was very high.) We would thus have very high confidence that Bronze Age sites were substantially larger on average than Neolithic ones. But it is not easy to say exactly what this means in terms of samples and populations. Literally, we have concluded that it is extremely likely that our sample of Bronze Age sites came from a population of larger sites than our sample of Neolithic sites. But those two populations may really be imaginary. If our survey was so effective that unstudied sites within the region are virtually nonexistent, then the population within the region is not substantially different from our sample. It may not be any more meaningful to think of this population existing outside the region either, since what was going on in the next region may well be entirely different—perhaps there were no Neolithic sites at all.

The population we have sampled in a case like this is truly imaginary. Thinking of the things we study as samples from imaginary populations may not sound like a very good way to approach reality, but it can indeed be meaningful to talk about significance and confidence even when the things we have studied do compose the entire population we are interested in. What an evaluation of significance like this last example tells us, in real-world terms, is that the quantity and character of the observations we have made gives us a real basis to discuss a change in mean site area from Neolithic to Bronze Age. The change we have observed is very unlikely to be due to the small number and equivocal nature of our observations. In short, we have enough information to say quite confidently that something changed and to proceed to consider more fully the nature of the changes and the forces that produced them. Such an indication enables us to put to rest one of the continual worries of the archaeologist: whether we have enough information to say anything at all. Whatever other doubts we may have about our conclusions, at least we do not need to worry that we did not recover information about enough sites to tell whether or not there was a change in site size between Neolithic and Bronze Age.

Returning to think about Early, Middle, and Late Archaic projectile points, suppose that we discover that the proportion of corner-notched points in the Early Archaic is 46% ± 23% at the 95% confidence level, and that the proportion of corner-notched points in the Middle Archaic is 34% ± 19% (also at the 95% confidence level). We would not be much interested in talking about what caused a change in the proportion of corner-notched points, because the quantity and character of the observations we have made does not make us very confident that there *was* a change. What we might be interested in doing is visiting more museums and observing more points. With a larger sample we would be able to achieve smaller error ranges at the 95% confidence level. Eventually we should be able to see either that there was a change that we could talk about with enough confidence to make the conversation worthwhile, or, alternatively, that whatever change occurred was so small as not to be very interesting.

In either of these two cases, application of the statistical notion of confidence (or its mirror image, significance) has told us whether the quantity and character of the observations we have made are sufficient to make some conclusions. Statistical reasoning provides us with powerful tools to deal with this concern. If you find that a difference between two sets of observations you have made has high statistical significance, then at least you know that you do not have to worry about not having enough observations on which to base conclusions. To say (as some have) that such a difference has high statistical significance, but that a larger sample is needed for us to be confident about it, is to reveal lack of comprehension of what the significance test means. At the very least, the high significance means that a larger number of observations is *not* needed.

What none of this touches is the problem of sampling bias discussed in Chapter 7. Having decided to treat the observations we have as if they were a random sample enables us to go ahead and utilize the tools of statistics. We may discover that our observations, from whatever they are really a sample, are an insufficient basis to find out what we would like to find out—that is, that whatever pattern we observe has very little significance. Or we may discover that the observations we have are sufficient to tell us some statistically highly significant things about whatever population it is that they are a sample from. When we arrive at this latter point, we are put in the position once again of considering what it is we have a sample from. If the things we observed were selected in a biased manner that affects the nature of the observation we are interested in, then we must reason our way around that problem as best we can. The assistance we get from statistics in that task is limited to helping us to delineate the problem of sampling bias clearly and, in a sense, compartmentalize it by separating it analytically from the issue of sample size, which the statistical tools in this book are designed to deal with directly.

Radiocarbon dating provides a context in which archaeologists are accustomed to reasoning in this way, although we do not usually talk about it quite like this. When a sample of carbon is submitted to a dating laboratory that tells us its age is 800 ± 100 years, we say that there is about a 66% chance that the sample is between 700 and 900 years old (since by convention the error ranges expressed with radiocarbon dates are one standard error). Before we can use that result to conclude anything at all about the date of a particular stratum, there are several other issues to consider. How confident are we that the sample was uncontaminated? That it did not fall down a rodent burrow from a more recent stratum? That it was not simply a burned root of a much more recent tree? That it was not from the long-dead heartwood of a tree already ancient at the time it was deposited? In effect, all these additional questions are questions about the real nature of the population of carbon atoms of which those counted in the laboratory were a sample.

The error range tells us whether the observations are sufficient to tell us what we need to know. Suppose that what we need to know is whether or not the stratum in question is more than 400 years old. A radiocarbon age of 800 ± 100 years gives us high confidence that the sample dated is more than 400 years old. Submitting a larger chunk of carbon to the laboratory in the hope of narrowing the error range would not be at all helpful. Increasing the size of the sample would be a waste of time and money. We are quite confident that those carbon atoms were a sample from a population of carbon atoms assembled more than 400 years ago. The sample was of quite sufficient size to tell us that. The worries about how the sample was selected (in effect, about sampling bias), however, remain. Larger samples and more statistics will not help us resolve those worries. We must reason our way through those difficulties as best we can with recourse to other considerations than statistics.

The approach taken in this book to the problem of sampling bias is similar to what has become customarily accepted practice in handling radiocarbon dates. Instead of treating random sample selection as a criterion that must be met before using the tools of statistics (which, if taken literally, would stop us dead in our tracks in almost every potential application of statistics to archaeology), the issue of what our observations are really a sample from becomes a consideration in what we can conclude from our observations—if they have enough statistical significance to make it possible to conclude much from them at all. If our observations do not have enough statistical significance to make them meaningful, then we do not waste time addressing the issue of bias in the process of sample selection. What we want instead is another (larger) sample, selected in as unbiased a manner as possible.

Making conclusions or interpretations, as discussed in Chapter 7, carries us beyond the realm of statistics, although statistical tools can help put us in better position to make conclusions. The significance or confidence levels we arrive at with statistical tools have to do with the probable nature of the populations (real or imaginary) that the things we observe are samples from. Knowing how much confidence to place in our observations (or equivalently, how much significance they have) helps us evaluate just how much support they provide for the conclusions we are interested in making. The confidence probabilities, however, do not apply directly to the conclusions.

If we think that family size increased from one period to the next and that such a change might result in larger storage jars in the second period, we can use statistical tools to determine how much confidence we should have in our observation that storage jar volume increased. If we find that we are 99% confident that storage jar volume increased, that does not mean that we are 99% confident that family size increased. There are other possible reasons for the increase in storage jar volume. The confidence probabilities pertain to our observations, not to what we think our observations mean. The high confidence we have in our observation of storage jar volume increase, of course, provides evidence in favor of our conclusion. (If our calculations gave us less confidence that storage jar volume increased, then our observations would provide less support for our conclusion that family size increased.) Observations at high confidence levels of other kinds of evidence consistent with increases in family size would add more support to our conclusion, while observations at high confidence levels of other kinds of evidence inconsistent with increases in family size would make us doubt our conclusion. The weighing together of these multiple, quite possibly contradictory, observations of different lines of evidence is essential to the process of evaluating conclusions, and it is not primarily a statistical task.

Statistical tools are more useful at a lower level in helping us to evaluate each individual line of evidence and to assess just how much (or how little) support each set of observations contributes toward sustaining our ultimate conclusions. It is because we use statistics most often in a context like this rather than in a context where we must make yes or no decisions based on observation of a single variable in a single sample that a scalar rather than null hypothesis testing approach to framing significance tests is particularly suitable in archaeology. Using samples simply to make estimates with error ranges about the populations they were selected from and comparing these estimates with bullet graphs is, in many instances, even clearer and more direct.

In particular, the statistical tools most discussed in this book provide a powerful approach to the archaeologist's perennial worry about whether there are enough observations to conclude anything much. Are the 253 sherds collected from this site enough to enable us to talk very confidently about proportions of different types? How confidently can we discuss the size of

Formative houses on the basis of the five house floors we have excavated? Do we have much confidence that the number of temple mounds depends on site area, given that our observation of this relationship is based on only 16 sites? Do we need to analyze all the unifacial flake tools recovered from this site, or could we learn more by studying a sample intensively? If a sample, how large would it need to be to tell us what we need to know? These are all examples of questions that have loomed large in archaeologists' sleepless nights for decades. They have been answered too often in archaeological research in purely arbitrary and subjective ways. Answering such questions on the basis of subjective impressions or "gut feelings" is unsupportable. It is precisely these kinds of questions that the statistical tools explored in this book can help us whittle down to size.

Suggested Reading

Bibliographic citations have been avoided in this book in order to streamline the presentation and because careful tracing of the intellectual pedigree of many of the ideas and techniques discussed here is a scholarly endeavor in itself, and one not comfortably combined with an introduction to their application in archaeology. The books and articles listed below, however, are places to go for further information on statistics in archaeology. The literature on statistics in archaeology has become very large, and the list below is both very short and slanted toward statistics in general more than toward statistics specifically in archaeology. Consequently, a large number of perfectly relevant references have not been included—the selection is idiosyncratic rather than comprehensive. Some of the items included are relatively new; some are not so new. Some are included because they share the general outlook of this book (and indeed in some cases are the specific inspiration for it); some, because they complement it (which is to say they take a different perspective).

GENERAL STATISTICS BOOKS

Exploratory Data Analysis, by John W. Tukey (Reading, MA: Addison-Wesley, 1977, 688 pages), is one of the classic presentations of an approach to statistics from which much in this book is derived, and its author is the father of the approach. Not surprisingly, there is a great deal more to exploratory data analysis (EDA) than has been presented in this volume, and readers who would like to go directly to the source to find out about it should read Tukey's book, which is a full-scale introductory text in EDA. Although EDA is now more than 20 years old, only parts of the prescription Tukey laid out for EDA have been much applied in archaeology (and even those parts that have been do not yet constitute the "standard" statistical approach in the archaeological literature). Many of the techniques Tukey discusses in his book were intended to be easily accomplished with pencil and paper or, at most, with a calculator, but more widespread availability in the most commonly used computer statpacks would undoubtedly encourage greater use of EDA techniques in archaeology and in other fields.

Exploratory Data Analysis, by Frederick Hartwig and Brian E. Dearing (Beverly Hills, CA: Sage Publications, 1979, 83 pages), is a brief presentation of the basic techniques of EDA. It nevertheless includes a number of EDA topics not covered in this volume.

Applications, Basics, and Computing of Exploratory Data Analysis, by Paul F. Velleman and David C. Hoaglin (Boston: Duxbury Press, 1981, 354 pages), is yet another introduction to EDA techniques, less formidable than Tukey's and more comprehensive than Hartwig and Dearing's.

Understanding Data, by Bonnie H. Erikson and T. A. Nosanchuk (Toronto: McGraw-Hill Ryerson, 1977, 388 pages), is an introductory statistics text that combines EDA with more traditional statistical approaches. It advocates two different and complementary kinds of work with numbers (exploratory and confirmatory), keeping the two strongly separated and emphasizing the differences between their goals. The presentation is especially accessible and free of jargon and abstract mathematics.

Introduction to Contemporary Statistical Methods, by Lambert H. Koopmans (Boston: Duxbury Press, 1987, 683 pages), also combines EDA with more traditional statistics. A very wide range of methods is covered, and the logic behind the methods is presented in more abstract mathematical terms than in most of the other books listed here. Instead of focusing on the difference between exploration and confirmation throughout the book, Koopmans considers statistical exploration at the beginning, and then complements the discussion of the usual significance testing techniques with a wide array of robust techniques suitable for use on data that present problems for the usual techniques.

Nonparametric Statistics for the Behavioral Sciences, by Sidney Siegel (New York: McGraw-Hill, 1956, 312 pages), is a classic presentation of a full array of robust techniques for evaluating significance, that is, ones that are not much affected by things like very asymmetrically shaped batches for which means and standard deviations are not useful. Many of these techniques require special tables in which to look up the results, and Siegel provides them.

Sampling Techniques, by William G. Cochran (New York: John Wiley & Sons, 1977, 428 pages), describes itself (quite accurately) as "a comprehensive account of sampling theory." It is, perhaps, the ultimate source on this subject. Estimating means and proportions, sample selection, stratified sampling, cluster sampling, sampling with and without replacement, determining necessary sample size, and many other topics are covered in detail. The full logic behind the techniques presented is given in mathematical terms.

Elementary Survey Sampling, by Richard L. Scheaffer, William Mendenhall, and Lyman Ott (Boston: Duxbury Press, 1986, 324 pages), covers much of the same ground that Cochran does. The presentation is largely in terms of abstract mathematics, but it is considerably less detailed and formidable than Cochran's.

INTRODUCTIONS TO STATISTICS FOR
(AND OFTEN BY) ARCHAEOLOGISTS

Refiguring Anthropology: First Principles of Probability and Statistics, by David Hurst Thomas (Prospect Heights, IL: Waveland Press, 1986, 532 pages), is an introductory statistics text specifically for anthropologists (including archaeologists). The approach is purely traditional (that is, it does not incorporate an EDA perspective or techniques), and some rules are laid down that this volume has argued against, but numerous robust methods are discussed. There are abundant examples of the application of all the techniques presented to real data from archaeology, cultural anthropology, and biological anthropology.

Digging Numbers: Elementary Statistics for Archaeologists, by Mike Fletcher and Gary R. Lock (Oxford: Oxford University Committee for Archaeology, 1991, 187 pages), applies basic statistical techniques (both traditional and EDA) specifically to archaeology. The presentation is informal, avoids jargon, and is designed to be very accessible, especially to those suffering math anxiety.

"Regional Sampling in Archaeological Survey: The Statistical Perspective," by Jack D. Nance (*Advances in Archaeological Method and Theory* 6:289–356, New York: Academic Press, 1983), attempts a comprehensive discussion of spatially-based sampling in the context of regional survey, although many of the same issues arise at smaller spatial scales as well. The statistical techniques involved in sample selection and in making estimates about the population are described, with special emphasis on cluster sampling. Numerous concrete and practical problems in archaeology are discussed with both hypothetical and real examples.

Quantifying Archaeology, by Stephen Shennan (Edinburgh: Edinburgh University Press and San Diego, CA: Academic Press, 1988, 364 pages), is an introductory statistics text (and more) specifically for archaeologists. Mostly traditional statistical methods are covered, but some EDA techniques are also included. Shennan goes beyond basic statistical principles to deal with multivariate analysis (with emphasis on multiple regression, clustering, and principal components and factor analysis). Methods for estimating population means and proportions are presented (not usual in introductory statistics books) and the special issues that sampling raises in archaeology are discussed.

Exploratory Multivariate Analysis in Archaeology, by M. J. Baxter (Edinburgh: Edinburgh University Press, 1994, 307 pages), contains only a very brief review of basic statistical techniques (including the fundamental ones from EDA). Its aim is to consider more advanced topics in multivariate analysis—techniques that can deal simultaneously with the patterns of relationships among numerous variables. Extended treatment is given to principal component analysis, correspondence analysis, cluster analysis, and discriminant analysis, and numerous examples of multivariate analyses of real archaeological data are woven into the explanations.

ARCHAEOLOGISTS CONSIDER STATISTICS IN OUR DISCIPLINE

"The Trouble with Significance Tests and What We Can Do about It," by George L. Cowgill (*American Antiquity* 42:350–368, 1977), makes the case for an attitude about significance testing that has inspired much in the perspective taken on this subject in this volume. It is a distinctly different view than is often adopted in introductory statistics texts—indeed it is branded as heresy by the rules often found in introductory statistics texts. This article is fundamental for those interested in a fuller presentation of the arguments that archaeologists will often find it useful to use samples directly to make estimates about populations and that it is usually a mistake for archaeologists to force significance tests into the mold of a yes-or-no decision. Cowgill's suggestions about the most useful ways to approach these issues in archaeology go well beyond what is presented in this volume, which has stopped at the point where the information commonly provided by computer statpacks imposes a limitation.

"A Selection of Samplers: Comments on Archaeo-statistics," by George L. Cowgill [In *Sampling in Archaeology*, edited by James W. Mueller (Tucson: University of Arizona Press, 1975, pp. 258–274)], prefigures some of the issues Cowgill argues more fully in his later paper (above), and focuses especially on sampling, criticizing many of what he sees as erroneous notions that appear in other papers in the same volume.

"On the Structure of Archaeological Data," by Mark S. Aldenderfer [In *Quantitative Research in Archaeology: Progress and Prospects*, edited by Mark S. Aldenderfer (Newbury Park, CA: Sage Publications, 1987, pp. 89–113)], is a discussion of the fundamental nature of data in archaeology, the position that numbers occupy in such data, and the implications that this has for how we think about and analyze data. The other articles listed below (and other interesting ones as well) all appear in this same volume.

"Quantitative Methods Designed for Archaeological Problems," by Keith W. Kintigh (in Aldenderfer, pp. 126–134), discusses the issue of the extent to which standard statistical techniques and those borrowed directly from other disciplines are suited to the particular needs of archaeology.

"Simple Statistics," by Robert Whallon (in Aldenderfer, pp. 135–150), stresses the importance of exploring the patterns in numbers in batches before proceeding to more complicated analyses.

"Archaeological Theory and Statistical Methods: Discordance, Resolution, and New Directions," by Dwight W. Read, and "Removing Discordance from Quantitative Analysis," by Christopher Carr (in Aldenderfer, pp. 151–243), try to place archaeological data analysis firmly in a broader context. Both authors are concerned that data analysis is too often conceived and carried out in isolation from the theoretical questions that analysis aims to help answer. As a consequence, "discordance" between data, analysis, and theory arises and seriously impedes the archaeological endeavor.

Index

INTERDISCIPLINARY CONTRIBUTIONS TO ARCHAEOLOGY
Chronological Listing of Volumes

THE PLEISTOCENE OLD WORLD
Regional Perspectives
Edited by Olga Soffer

HOLOCENE HUMAN ECOLOGY IN NORTHEASTERN NORTH AMERICA
Edited by George P. Nicholas

ECOLOGY AND HUMAN ORGANIZATION ON THE GREAT PLAINS
Douglas B. Bamforth

THE INTERPRETATION OF ARCHAEOLOGICAL SPATIAL PATTERNING
Edited by Ellen M. Kroll and T. Douglas Price

HUNTER–GATHERERS
Archaeological and Evolutionary Theory
Robert L. Bettinger

RESOURCES, POWER, AND INTERREGIONAL INTERACTION
Edited by Edward M. Schortman and Patricia A. Urban

POTTERY FUNCTION
A Use-Alteration Perspective
James M. Skibo

SPACE, TIME, AND ARCHAEOLOGICAL LANDSCAPES
Edited by Jacqueline Rossignol and LuAnn Wandsnider

ETHNOHISTORY AND ARCHAEOLOGY
Approaches to Postcontact Change in the Americas
Edited by J. Daniel Rogers and Samuel M. Wilson

THE AMERICAN SOUTHWEST AND MESOAMERICA
Systems of Prehistoric Exchange
Edited by Jonathon E. Ericson and Timothy G. Baugh

FROM KOSTENKI TO CLOVIS
Upper Paleolithic–Paleo-Indian Adaptations
Edited by Olga Soffer and N. D. Praslov

EARLY HUNTER–GATHERERS OF THE CALIFORNIA COAST
Jon M. Erlandson

HOUSES AND HOUSEHOLDS
A Comparative Study
Richard E. Blanton

THE ARCHAEOLOGY OF GENDER
Separating the Spheres in Urban America
Diana diZerega Wall

ORIGINS OF ANATOMICALLY MODERN HUMANS
Edited by Matthew H. Nitecki and Doris V. Nitecki

PREHISTORIC EXCHANGE SYSTEMS IN NORTH AMERICA
Edited by Timothy G. Baugh and Jonathon E. Ericson

STYLE, SOCIETY, AND PERSON
Archaeological and Ethnological Perspectives
Edited by Christopher Carr and Jill E. Neitzel

REGIONAL APPROACHES TO MORTUARY ANALYSIS
Edited by Lane Anderson Beck

DIVERSITY AND COMPLEXITY IN PREHISTORIC MARITIME SOCIETIES
A Gulf of Maine Perspective
Bruce J. Bourque

CHESAPEAKE PREHISTORY
Old Traditions, New Directions
Richard J. Dent, Jr.

PREHISTORIC CULTURAL ECOLOGY AND EVOLUTION
Insights from Southern Jordan
Donald O. Henry

STONE TOOLS
Theoretical Insights into Human Prehistory
Edited by George H. Odell

THE ARCHAEOLOGY OF WEALTH
Consumer Behavior in English America
James G. Gibb

STATISTICS FOR ARCHAEOLOGISTS
A Commonsense Approach
Robert D. Drennan

DARWINIAN ARCHAEOLOGIES
Edited by Herbert Donald Graham Maschner

CASE STUDIES IN ENVIRONMENTAL ARCHAEOLOGY
Edited by Elizabeth J. Reitz, Lee A. Newsom, and Sylvia J. Scudder

HUMANS AT THE END OF THE ICE AGE
The Archaeology of the Pleistocene–Holocene Transition
Edited by Lawrence Guy Straus, Berit Valentin Eriksen, Jon M. Erlandson, and David R. Yesner